Cowgirl Sass and Savvy

© 2007 Julie Carter
julie-carter.com

978-0-9741627-5-1

PricePoint+Publications
in cooperation with author

Republished in 2016

COVER ART by Gordon Snidow, from his American Woman Series, titled, "If Momma Ain't Happy, Ain't Nobody Happy."

Snidow has been known as the foremost chronicler of the contemporary cowboy since 1959. He is a Charter and Emeritus member of the Cowboy Artists of America. For almost 50 years, Snidow has created and painted his own chronicle of the modern West.

See the Art of Gordon Snidow at http://www.gordonsnidow.com

Table of Contents

Sass & Savvy from the Ranch

Cowboys...You Gotta Love 'em

Rodeo and Ropings

Cowgirl Philosophy

Down Dusty Roads

Rural Holidays

The West is dead, my friend
But writers hold the seed
And what they sow
Will live and grow
Again to those who read.

– C.M. Russell, 1917

FORWORD:

Like C.M. Russell, I believe the West will continue to stay alive in the words of those who paint word pictures.

This book has been a work in progress for more than four years, as each week God's gift of inspiration gave me a new idea, a new story. Without the gift, without the inspiration, there would be no book.

My stories come from the heart, come from my memory and come from the people in my world. They are stories about growing up on a ranch in the mountains of southern Colorado, of my experiences on the rodeo road and more ranch living and cowboy life in New Mexico.

I am blessed with a number of friends who share their ranch-wife, ranch-life stories with me, knowing they will end up in print at some point. These are bright, beautiful, strong, independent women, each with a great sense of humor.

Ranch life creates within each of us an ability to view the world from a different angle, usually from the bottom looking up. It sorts out priorities on a survival rating scale. We don't have much time to spend worrying about things that don't really matter in the big picture of life.

Our survival gear includes a Bible and the ability to find something to be thankful for daily. Most days we soothe our souls with laughter, and the best of that is when we laugh at ourselves.

We have no patience for whiners, but we will spend our last breath helping someone truly in need. We get up early each day ready to tackle the next

24 hours, one situation at a time. Rarely do our situations come in single file, but we "cowgirl up" and get through it.

We doctor children, pets, livestock and husbands. We have bottle fed babies, birds, puppies, kittens, rabbits, calves, colts and the occasional fawn. We mend fences and torn britches.

We live in places where the directions to get there include the words *miles*, *last*, *cattle guard* and *dirt road*. Our early morning prayers cover rain, cattle, family, friends and an assortment of 4-H projects that will include a goat or a pig.

It is the norm to find our conversations far from the usual "woman talk" of hair color, the latest fashion in pumps and purses or a Saturday night concert. We learn about that stuff from the magazines.

Most often we talk about feeding cattle, calving heifers, a new baby colt, the veterinarian's last visit, and sometimes the location of the next county fair pig sale. Of course there is always the moment given to a story about "What *was* he thinking?"

Occasionally we share funny books we have read and movies we liked. Potluck recipes are a given as often as the brand name for the most comfortable work boot.

We understand each other. We know that fixing supper and fixing the stock tank float is not an either/or choice. A knowing nod always follows the first line of a story that begins, "We went to check a gate and he said we'd be right back."

It is for these women and the thousands like them that I write these stories, putting their life on paper. I write for those pioneers who came before us and for those that, with fresh excitement, are now embarking on a first adventure in ranch life. I write for those who don't write, but whose lives are the very foundation of my words.

May we always find a way to laugh together and may these stories gift you with a moment of joy at a time you most need it.

Julie Carter

DEDICATION

This book is dedicated to my family and friends who have provided continued encouragement to me, not just for this project, but also for life in general. I thank God for answered prayers and His promise to never let me go.

Sass & Savvy from the Ranch

Ranch Wife 101

My first idea was to create a list of rules for cowboys written by cowgirls – things that would bridge the understanding gap. But historical reaction of cowboys to advice given by their wifely partners made me realize the futility of that effort.

Moving from that fleeting moment of "saving the cowboy world," I decided to help a small part of it by suggesting some basic advice to those considering matrimony to a cowboy. While this is by no means a complete guide, recognizing the following situations will save years of misunderstanding.

Ranch wife 101 guidelines:

1. Always load your horse last in the trailer so it is the first one unloaded. By the time he's got his horse unloaded, you will have your cinch pulled and be mounted up ready to go – lessening the chance of him riding off without you with your horse trying to follow while you are still trying to get your foot in the stirrup.

2. Never, and I repeat, never, ever believe the phrase "We'll be right back." When he has asked you to help him do something out on the ranch, those echoing words, "this will only take a little while" should evoke sincere distrust in every woman who hears them. The only promise in those words is you won't be back until long after dark.

3. Always know there is absolutely no romantic intention when he pleadingly asks you to take a ride in the pickup with him to go around the ranch while he checks waters and looks at cattle. What that sweet request really means is – he wants someone to open the gates.

4. He will <u>always</u> expect you to quickly be able to find one stray yearling in a 4-section brush-covered pasture, but he

will <u>never</u> be able to find the mayonnaise jar in 4 square feet of refrigerator.

5. Count every head of every thing you see – cattle especially, but sometimes horses, deer, quail or anything that moves. Count it in the gate, out the gate or on the horizon. The first time you don't count is when he will have expected that you did. That eyelash-batting blank look you give him when he asks, "How many?" will never be acceptable to him.

6. Know that you will never be able to ride a horse or drive a pickup to suit him. Given the choice of jobs, choose throwing the feed off the back of the pickup. If he is on the back and you are driving, the opportunity for constant criticism of speed, ability and your eyesight will be utilized to the full extent. "How in the *@*# could you NOT see that hole?" he will demand.

7. Never let yourself be on foot in the alley when he is sorting cattle horseback. When he has shoved 20 head of running, bucking, kicking yearlings at you and then hollers "Hold'em, hold'em," at the top of his lungs, don't think that you really can do it without loss of life or limb. Contrary to what he will lead you to believe, firing yourself and walking back to the house is always an option.

8. Don't expect him to correctly close the snap-on tops on the plastic refrigerator containers, but know he will expect you to always close every gate, always. His reasoning, the cows will get out, the food will not.

9. Always praise him when he helps in the kitchen – the very same way he does when you help with the ranch work – or not.

10. Know that when you step out of the house you move from the "wife" department to "hired hand" status. Although the word "hired" indicates there will be a paycheck, which you will never see, rest assured you will have job security. The price is just right. And most of the time you will be "the best help he has," because you are the ONLY help he has.

The Art of Plotting

Ranch girls are raised to "make a hand" at whatever it is they do. Ranch girls that become ranch women use those formative years to hone their many skills, and one of them becomes the fine art of "plotting."

There is just no other way to say it. Working with a husband at the ranch can be pure hell. After about three decades of never being able to do it right, the Little Pretty will begin plotting miniscule revenges.

A number of phrases will be completely eliminated from her vocabulary. Never again will you hear her say, "Nice home-cooked meal," or "Oh, you shouldn't have done that for me." And absolutely, she will never again utter, "No, I don't mind driving those 50 miles out of the way to pick that up for you."

A unique sign language develops between the couple over the years. The unspoken instruction that passes between them without much notice accomplishes plenty in a day's time.

So when he stands up in his saddle and hollers obnoxious instructions at her, actually meant for the cowboy riding on the other side of her that he doesn't want to offend but he wants to hear it, her sign language takes on gestures better not put into words by a lady.

Knowing he will always cannibalize her household tools and her horse tack requires extra measures for her to protect both. Turning her personal tools at the house to a nice hot-pink shade with a can of spray paint seemed to slow the trend of them "going missing." That "little girl" toolbox just isn't worth stealing.

While it seemed a little extreme, after losing several good headstalls and a rope or two to his "Oh, I'll just take hers," mindset, hiding what she wanted to keep under the mattress worked like a charm. It went to the barn only when she did.

Sorting cattle in an alley can be, and most often is, perfect grounds for a divorce. Some women get it figured out early in the game so they don't get frequent invitations to help.

Self-proclaimed "pipe dealer to the stars," calling himself that ever since he sold a load of pipe to world champion team roper Speed Williams, tells about sorting cattle with his wife.

He says to her, "Let the Charolais by."

"Wait! Stop everything," she says. "Do you mean the beige one?"

"Yes, the beige one," he tells her. "Now let the paint in."

"Wait! Stop everything," she says again. "You mean the plaid one?"

He said he had to relate everything in sewing terms and now he only sorts cattle with married men who might actually understand his terminology.

The veteran cowgirl who missed the "play dumb" plaid and beige lesson will get her usual assignment. It will be something like "Here, stand directly behind this gate and hold it so the cows don't break it down."

And he will never understand why, that when he says "I sure could use some help," her answer will be, "I'll do my very level best to find you some."

Vacation Options for the Ranch Wife

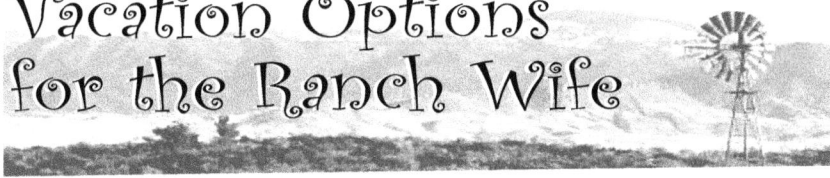

Ranchers have a calendar marked full of red-letter dates from November to March, which he will call "vacation options for the wife."

The schedule for a "holiday off the ranch" is specifically targeted to fit between fall shipping and spring calving. Actually getting off the ranch will depend on the weather and an assortment of other things that can and usually do go wrong.

With the dead of winter back-grounding these plans, the options always include some sort of livestock convention or like event with agriculture tags all over it. Destination titles read Stewards of the Range Conference, National Angus Convention, National Conference on Grazing Lands, National Auctioneers Winter Seminar and the state's annual livestock convention held in every state that has a cow.

The Little Pretty will pack her best prairie skirt, a few strands of silver and turquoise and maybe even get her roots touched up. Off she'll go with Livestock Man to stay in a fancy motel, eat at endless buffets, wait out never-ending meetings and hope she can make a dent in the fall cattle check at the nearest mall.

Sometimes getting that far away from home isn't possible for a change of pace and a dose of cabin fever medicine. So Livestock Man will take the Little Woman to a local happening which can often be as worldly as the 54th Annual Fall Testicle Festival or as simple as a community church supper.

Wives of competition ropers always get set up for an outing that is supposed to be for their pleasure but you can bet there will be a roping involved. The church ropings are an example of the blending of reverence with, "Hey, may as well, we're here anyway."

Sometimes the roping "wolves" will try to slip in and win the money away from those church ropers only to find that these holy ropers have been spending a little more time practicing than in church.

Men with livestock are a dedicated bunch. They have been known to postpone a funeral until after the Wednesday auction because during the fall, sale day has priority over life itself. There is no rush to get the deceased in the ground. He will still be dead on Thursday.

As a last resort, Livestock Man will take his bride to the all-you-can-eat catfish night at the local watering hole and actually call it a vacation. Hollering over the noise of the jukebox and peering through a smoky haze originating from the patrons and the kitchen, she knows this is the pinnacle of her winter fun.

The bartender, who never sold a sip he didn't first test with his own liver, offers some conversation amidst the clacking of the balls on the pool table as a few locals hold their own version of the world championships.

While he chats, she listens and occasionally ducks when the back end of a pool cue is aimed at her head during a strategic shot.

Wherever "the vacation" ends up this winter, know the bride of Livestock Man will soon be back mucking through melting snow to find a new baby calf. She'll be breaking ice on the drinkers on her half of the feed route and mentally planning for warmer days ahead. A day with temperatures above freezing will make her simply giddy.

As the saying goes, "if this ain't livin' then kick us all out."

Martha, Maxine and Me

I've picked on Martha Stewart before and she is probably used to it by now. Besides, right now she is busy maintaining her post-prison life so likely won't notice.

I'd like to mention a few things Martha never had to deal with like baby calf afterbirth under her nails and broken stock tank floats. I doubt she ever found herself having to wait on the hens to lay an egg so she could finish baking a batch of cookies.

I think that qualifies me to speak from a position of authority from my side of the cattle guard.

There is Martha's way and then there is my way, which to save a plagiarizing charge, also happens to be Maxine's way. You all know Maxine, that crabby character from Hallmark that is so very witty, flippant and the irreverent lady of one-liners. She's my hero. I like the way she thinks.

Maxine has a way of cutting to the bottom line. A real shoot-from-the-hip kinda gal.

A few kitchen hints from Martha invoked Maxine's brazen honesty.

- When a cake recipe calls for flouring a baking pan, use a bit of dry cake mix instead and there won't be any white residue on the outside of the cake. To which Maxine replies: "Go to the bakery! They'll even decorate it for you."
- To keep potatoes from budding, place an apple in the bag with them. Maxine's hint is: Buy Hungry Jack mashed potato mix and keep it in the pantry for up to a year.

Maxine's sound logic is painfully honest and laughably correct:

- The trouble with bucket seats is not everybody has the same size bucket.
- Drinking makes some husbands see double and feel single.
- Living in a nudist colony takes all the fun out of Halloween.
- After a certain age, if you don't wake up aching in every joint, you are probably dead.

Every now and then we rural women get an urge to be more like Martha and less like Maxine. Fortunately, we recover from that lapse in judgment fairly quickly.

A ranch wife friend said she changed hair stylists, which meant she went from a beautician to a stylist. She had been told a stylist was a much higher-class person and while she is still not so sure the class was higher, the price certainly was.

Since no one seemed to be appreciating her "inner beauty" she was making an attempt to improve the outer version. After getting "styled," she said the grocery stop at Wal-Mart was very basic in an attempt to put some balance back into her household budget.

When I grow up I want to be just like Maxine. Some will tell you I am already dangerously close. I have never, ever been accused of being like Martha. Although, when Martha got out of jail last year it was reported she went to "her large ranch 40 minutes from New York." All this time I thought the closest thing to "ranch" Martha ever came close to was her homemade ranch dressing.

It could happen – perhaps we, Martha and I, could go out in the garden and see if those bulbs I didn't plant didn't come up. Or better yet, Martha and I would go riding off into the sunset with Maxine snidely commenting about the size of our "buckets" in the saddles.

Cowgirl Glamour is Hard Won

My first clue that life was not as it had been was when my subscription to *Glamour* magazine expired and was replaced by *Country Woman*. My mother intended it as a gift, but I know now it was warning.

After many years out in the workaday world offered by various major metro areas, I found myself "back at the ranch." That was more than a decade ago. I vowed I would not let myself fade into the rough and dust of the hard rural lifestyle. I refused to leave my femininity at the cattle guard even though many of my "new" jobs would be less than girly. Let me just say, it has not been easy.

One of my efforts at chic, out here in the middle of nowhere, is to have my nails done occasionally. You know, in the acrylic version, the kind of long nails that no one I know can really grow on their own. During the holidays or for special events, it's such a nice touch and I feel so elegant. I know, it's a girl thing.

Then came calving season. The first calf I pulled, not one nail broke. I was so proud! An hour later my elation turned to resigned despair. The amniotic fluid had eaten the acrylic like acid. The nails were gone.

I also discovered that the wormer we use on the cattle and horses does the same thing. No wonder it kills worms.

My manicurist couldn't decide if she was amazed or nauseous. Not wanting to get "that look" from the boss, my husband, I didn't mention what had happened. He thought the nails were a "dumb idea" anyway. He took no pride in me being the only "cowboy" help with really nice nails.

For those days that I am volunteered to go to "town" (population 1,000), I dress up. That means nicer jeans and a clean shirt. I do change my hairstyle from "birds nest under a ball cap" to "the wind always blows on a good hair day" just for the trip.

The family "car" is a 4-wheel-drive pickup. The best part of this high-built vehicle is you can depend on it. Dependably, every time you crawl down out of it, the backside of your pant legs will be covered in dust and dirt.

I make regular attempts at testing to the fullest the finest in moisturizers and skin care. I will never stop wanting to be soft and smell good. I work very hard at being an integral part of this life, but I refuse to fade into the droughted-out countryside.

I can rope a cow, pull a calf and work an alley gate with the best of them. I can fry whatever it is they want fried and make good gut-stickin' gravy.

I've come to expect the horses to blow snot on my clean white shirt when I'm dumb enough to get that close to them in my town clothes. I have even accepted being compared to a cow in the terms of being "ringy" when the wind blows for months on end.

I love the freedom of this life and all it offers in day to day pleasures. But, there is one thing I'd request if it would not fall on deaf ears: I would just like a little respect when I finally put my foot down and say, "No, I can't do that. I might break a nail!"

Terms of Endearment for the Biscuit and Gravy Boys

It was bound to happen eventually but it still surprised me. Somewhere out there in reader-land is an uninformed man I offended with the use of the term "redneck."

Somehow he had missed enough of my stories to not realize I claimed a redneck heritage for myself without having to chew "baccy" and have a big enough space between my front teeth to pick them with a fence post.

He kindly educated me on rednecks and how they were poor whites with a long history of being put down and slandered. And he never mentioned duct tape, pickup trucks or coon hunting with a pack of hounds.

He called me a bigot and told me I was hurtful and hateful. So I wrote him a "nice" letter and told him he was very wrong about me and seemed to have little understanding of rednecks.

I explained to him that my particular use of the word was a term of endearment for those that prefer biscuits and gravy over grapefruit and grapes.

Although rednecks will claim original roots in the pre-civil war South with a Confederate flag waving overhead, today they are everywhere. They have families, hold down jobs and make pretty decent neighbors. They are good people who have a strong sense of right and wrong and won't change their ways just because someone said they should.

The redneck ends up the butt of many jokes but doesn't care and in fact laughs the hardest. He is the first to help someone in need but the last to ask for help himself.

He wears many hats. He may be an executive in a suit. Yes, rednecks have crossed over to a white collar set. He could be a truck driver, bartender, farmer, cowboy, sailor, Green Beret, computer tech., teacher, waiter or airline pilot. And, sometimes he's even the President of the United States.

I use the "R" word on a regular basis with complete understanding of whom I speak. I take Martha Stewart's advice for rednecks on a regular basis, although I don't always agree. Livestock is not always a bad choice for a wedding gift.

I know that "y'all" is singular and "all y'all" is plural. I hear and use the term "fixin' to" almost daily and I am among those that have been excused from school because the cows got out. I know exactly what calf fries are and eat them anyway. I agree that he should offer to bait your hook for you, at least on the first date.

You will never hear a redneck say, "I'll have the arugula and radicchio salad." "Would you like your fish poached or broiled?" Or "Honey, did you mail that donation to Greenpeace?"

I'm not at all surprised when I find movie rentals, ammunition and bait all in the same store. And I do know someone who used a football schedule to plan their wedding date. Those beloved rednecks open doors for women without prejudice to age and everyone in a pickup waves at the vehicles they meet. It is called being friendly.

I can only hope that my uneducated "redneck media monitor" reads this story. I want to thank him for setting me on a mission to share things about the people who will always step up and say, "Yep, I'm a redneck and proud of it."

Axe-Handle Reputations are Made in the Line of Duty

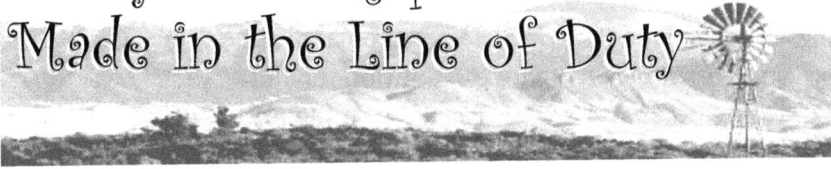

Reputations are made in a number of ways. Some are established over a lifetime of works, good or bad, and sometimes the distinction will come from a single moment in time without real word or surmised deed.

The bitter cold that has gripped the country this week induced thoughts of those folks that daily handle an axe. Breaking ice on livestock waters and splitting firewood are a few basics in this world that remain the same today as they did a century or more ago.

Out in the panhandle of Texas stand thousands and thousands of cattle on wheat pasture in someone's care. And yes, this very morning, those folks are making their "circle" – breaking ice, checking hot fence wires and a given: doctoring sick yearlings. Sub-zero temperatures are no excuse to stay home by the fire.

A few years back a couple I know was well into a long winter of heavy snow with plenty of ice and mud. They had been in the wheat cattle pasture business for a number of years, had good reputations for being fair wheat traders, quality pasture managers and just being generally honest folk.

On this particular day, the Little Woman, in both name and size, had made her circle which involved 65 miles of road driving. It entailed riding through 18 pastures checking for sick cattle and breaking ice on the troughs in each pasture. She had one more to go. Her good roan horse had been in and out of the stock trailer so many times on that long day he decided to not be in a particular hurry to unload.

Her "tired" was hanging out as much as the roan's but the job was to be done. Standing on the fender of the trailer and reaching through the slats, the cowgirl used the handle end of her ice-breaking axe to try to snare the reins on the horse's neck. While she was encouraging him to back out of the trailer, the farmer who owned that particular wheat pasture land happened by.

The cowgirl had the business end of the axe in her gloved hands, but the farmer was unable to see that. He slowed his pickup to view the action then quickly sped away as fast as the snow and mud would allow. Not giving it a thought, she got the horse unloaded and went about her business.

The next morning, midway through her circle, she stopped by the grain elevator to get a hot cup of coffee and thaw out by the ever-blazing stove. As she walked in, all the domino players stood up. Ordinarily she would only get a nod and a hello from them, so the standing ovation caused her to ask what was up.

"Lyndon was by here yesterday and said he saw you hit your good horse in the head with an ax because he wouldn't unload," said one of the men gathered at the elevator. "Everybody knows how much you love that roan horse, so we figured if you are that tough, we better come to attention when you walk in."

The cowgirl just laughed at them and never told them any different. That little smidgen of fear could be used to every last bit of advantage over the coming years.

Cats and Rats...
a Country-Style Crisis

We don't fret much over ballet recitals, front row seats at the symphony or trying to keep track of opera glasses. Fine wine in crystal glasses rarely breach the barriers of the beer and burp set.

With fair regularity, one of the crises dealt with in rural America is the lazy cat and fat rat problem. There is something wrong when the weekly trip to the grocery store includes another round of cat food and additional rat bait for the barn.

Two recent stories from two different states tell me this is not a problem unique to any one region.

Her morning started quite peaceful. Just as she was about to pour her first cup of coffee, the pound cat – called such because of his origin – came through the open patio door with a prize for her. Completely indoctrinated to the plan that food comes in a can, but country cat enough to be self-sufficient, he seemed to be supplementing his breakfast with a rat.

The pound cat also seemed to think he needed to eat this rat in the kitchen where everyone else eats. The mega-rat was in the walking-wounded category having survived the play-torture phase common to cats that catch rats.

When the cat put the rat down, the Little Woman grabbed a dishtowel and made a mad dash for the rat, surprising even herself with her circumstantial speed.

About the time she had the rat captured in the towel, her resident cowboy appeared and said he needed to talk to her. She was already in rapid motion toward the door and didn't slow down at the sound of his voice. This caused a loud "STOP!" command from the cowboy and she froze instantly in place while the dog ran to hide under the bed.

Calmly the cowboy told her his earthshaking can't-wait-another-second message, "Don't put the roping cattle out today. I want to look at them."

She showed him her rat, who was actively objecting to being wrapped in a dishtowel. The cowboy was unimpressed and left to start his day. Shrugging, her thought was, "This day has nowhere to go but up."

Throwing the rat out, she gave the dish towel a decent burial, washed her hands a dozen times, mopped the floor and hoped the coffee Juan Valdez was promising would make things look a little brighter. Maybe the dog would even come out from under the bed soon.

The other story involves five fat lazy cats living on the porch, two more with babies in the storeroom and four "even stupider dogs."

Much to this gal's surprise, she sees a fat healthy rat on the outside window sill. Looking around, all the resident cats are asleep on the porch. She gathers up an armload of them and takes them to the rat. They look, stretch and purr. She even holds one up to the rat with the only response a swish of the cat's tail and a purr.

So with broom in hand, she swats the rat off the sill and he runs behind a big flower pot. The cats seem amused but never move. The rat runs up the wall and under the eaves. One more broom bat and its down again. The cats don't move.

However, the "ranch security" dogs arrive to save the day. They run the cats back to their perches and a huge cat and dog fight ensues. This makes the broom again a handy item to swat cats and beat the dogs off the cats.

The rat is still on the eaves. Another broom swish brings it down and it makes yet another dash to the flowerpot. The dogs chase after it and the flowerpot goes flying, sending potting soil everywhere.

The dogs chase, toss and play with the rat until someone takes pity on it. The dogs have had a grand time while the cats continued to sun themselves, lick their fur and give an occasional glance at the dogs.

Then there is the serial killer ranch cat. He kills not for food but for attention as he stacks his kill at the door for his mistress' approval.

Both ranch wives agree their cat food bill is about to take a serious drop.

A Rope and Prayer Evened Up the Score

When the sun set that Saturday the score was: Cow 1, Cowgirls 0. And, it didn't take all day for it get that way.

I'll start at the beginning. The herd of cows had been moved out of the pasture a few days earlier. One mama and her very big should-already-be-weaned "baby" were unintentionally left behind.

We unloaded the horses at the top side of a six-section pasture. She and her calf could be seen in the distance, already traveling in a high trot in the wrong direction. So we pulled our cinches, got aboard our trusty cow-horses and loped after her. We eased around the pair and got them turned and headed toward the nearest corrals.

Momma cow took it well. Not ever slowing down much, she and the calf and we were headed in the right direction. Several miles of long trotting and we were almost there. We figured the easy part was over but we proved it within the hour.

This cow – I use that term only for propriety because I can and did call her many other things – had absolutely no intention of going near the corrals let alone in them. With just two of us horseback, we did all we knew how to convince her she was surrounded. She wasn't fooled; not for a second.

After she made tracks right over the top of us three times in her determination to head back to the hills, we reassessed the situation.

The washed-out gullies, rocks and hillside looked like a war zone and was about as easy to maneuver through on a running horse. The horses were worn out and the cow wasn't.

All hope of having a real "cowgirls rule" kind of day was gone. We needed reinforcements and preferably ones that could rope.

I had considered the roping option during the battle but knew I was outgunned – one very big cow and one old woman with a rope riding a very tired horse is not a fair fight.

We reconvened after church on Sunday. Church was one of our better ideas in this whole thing. It definitely needed prayer. We had let that cow get away three times and that is a hanging offense in most cowboy circles.

And because of it, she was going to be considerably harder to gather the second day. She'd developed a really bad attitude.

Reinforcements came in the form of two cowboys with ropes. Things were looking up. It took quite awhile longer to find her and true to what I thought, she didn't let us drive her even as close to the corrals as she had the day before.

Her calf took off with his eyes glazed over and no convincing would turn him back. He soon wore a nylon necktie (the loop of a cowboy's rope), and was tied up to await the arrival of a trailer.

Momma cow refused any negotiations for a peaceful capture and soon wore not one but two ropes. Her desire to drag us off, horse and all, made the use of an additional rope an easy decision.

After some huffing, puffing and a few tense minutes, the pair was loaded in the trailer and hauled off to the cull pen at the headquarters.

Now before you cowboys get too puffed up over the fact we cowgirls needed some help, let me point out a couple things.

First, it took less than an hour for us to figure out we needed help. And second, we were smart enough to ask for it. There are many stories about how different that is from the "cowboy way."

A Kept Woman...
a Matter of Perception

Two women at a citified social gathering were having a conversation. The event was a collection of people who don't walk in the same world on a day to day basis. Those kinds of get-togethers usually prove that perceptions are not even close to reality.

"What do you do for a living?"

"I work at the ranch."

"Oh, so you don't have a job?"

"Yes, I work at the ranch."

The knowing "I see," that was spoken came with an unspoken attitude that said, "I'm talking to a 'kept' woman."

Oh yes, a kept woman. A ranch wife is indeed a kept woman. The list of her "kept" duties is endless.

She kept going to the corrals every night for three months in sub-zero temperatures to check on calving heifers.

She kept hay hauled to the barn and ice on the tanks broken.

She kept the fuel and feed suppliers on call so tanks weren't empty and the cattle didn't go hungry.

She kept the horses wormed, the horseshoer scheduled, the saddle house swept and the saddle blankets washed.

She kept the horse in the corral and the dogs out of the corral.

She kept the gate open while she waited for him to show up or she kept the gate closed because he didn't.

She kept the pantry full, hot food on the table, clean clothes in the drawer, and the ever-proverbial "home fires burning."

She kept the grocery list, the spare parts list, and the Christmas card list.

She kept the calendar marked with family dates, weather reports, cattle working dates and new calf tallies.

She kept the socials "social," his soul prayed for, and his mother's birthday remembered.

She kept the bills paid, his boots soled, and the sunscreen where he might think to use it.

She kept the coffee ever handy, the iced tea ever available and his grandfather's recipe for a hot toddy as a remedy for his achy body.

She kept track of where he put his stuff so he could find it.

She kept his butt out of a jam and the list is too long to say when and why.

She kept the vaccine guns washed and in one place and the vaccine front and center in the refrigerator.

She kept the kids fed, clean and in school and made sure he remembered which one was which and if he needed to say "happy birthday."

She kept her legs shaved and reminded him his face needed shaved.

She kept the clippers sharp for haircuts for him, the kids, the dog and the horse.

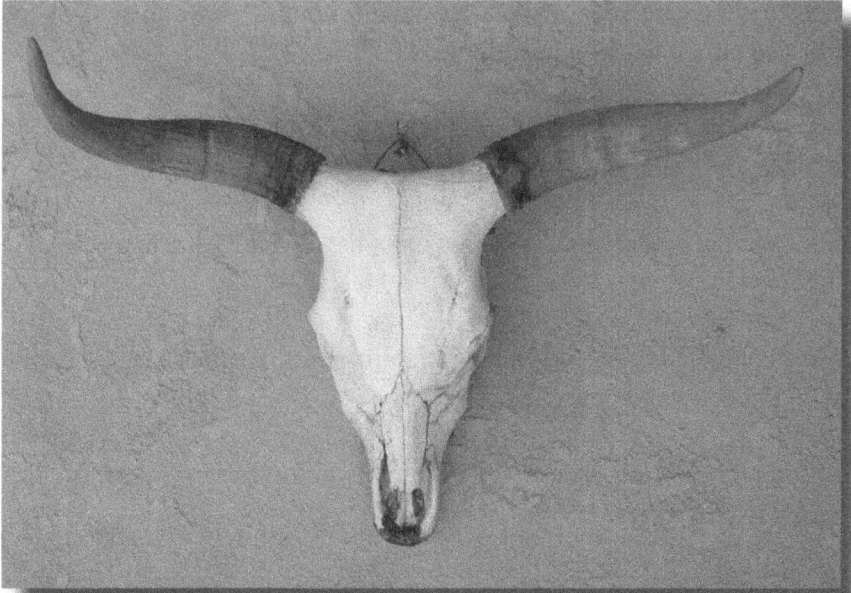

And through it all she kept her sense of humor even on days when humor was as scarce as rain.

Yes indeed, she is a kept woman.

In fact, she has a Ph.D. in "kept" and most days it would be measurably easier to have a real job.

However, as a rule, she would not trade this "kept" life for a two-story house in the suburbs with a new SUV in the driveway and hair and nail appointments every Friday followed by lunch at the club.

She is a kept woman because "kept" is who she is and what she does best.

Breaking in Tourists and City Kinfolk

Every now and then the head honcho on the outfit, while he's in town for supplies, will run into some adoring fan of American cowboy life.

The boss man will be picking up a new belt for the pump jack at the parts store, fueling up the pickup at the filling station or stocking up on ice at the grocery store and some total stranger will strike up a "gee, I've always admired cowboys" kind of conversation.

In addition, without knowing any better, the fan will suggest that perhaps he and his family could come out to the ranch as they are having a slow day in their vacation time. Overburdened with politeness, the boss man will tell them "Sure, come on out."

This same type of "invite yourself" happens most often with shirt-tail relatives who want to play cowboy, get a good meal and have some tall tales to tell back at the office on Monday.

They will insist on coming to "help" during a branding, shipping or any other fair-weather activity that might be going on at the ranch.

They never understand the amount of extra work they require just by showing up. Beside the room and board and necessary shepherding around the ranch, it's not unusual to have to call out a search party for one or more of them before the day is over.

An entire family will arrive, pour out of the mini-van and only occasionally will they offer up a bag of Fritos to add to the meal to be served later in the day.

Since the gentle horses usually belong to the Little Woman of the ranch, it is her horses that are assigned to the city kinfolk. These gentle beasts of burden are the only ones on the ranch safe enough for the dudes to fall off and not be walked on and kicked.

The boss won't cut Momma any slack when it comes pulling her share of the work, so she'll get to ride the green-broke colt and take the outside circle. "That colt's been needin' some miles anyway," he will convincingly tell her.

After starting lunch in the slow cookers to feed 30 people, Momma assured everyone the day could begin.

During one of these kinfolk' invasions, the young colt objected to getting pushed a little too hard when Momma was trying to be where she needed to be on time. The physical form of objection, called "watch me buck you off woman!" landed Momma on her backside.

As you would imagine, the colt took off for home at a dead run and Momma had to limp in on foot with her wrist hanging at a funny angle.

Somewhat short on sympathy, the Boss mentioned that old women shouldn't get bucked off. He even dared seem a little put out that he would now have to pick up the slack to cover for her lack of help.

After the branding was done and most of the Coors gone from the cooler, he got around to taking her to the emergency clinic.

For six weeks, hundreds of bologna sandwiches were eaten by the masses. Amazingly, it kept the kinfolk head count to a minimum. The boss, all the ranch hands and the kids swore they would never loan Momma's horses to the city kin again.

Not long after the cast came off, another crop of tourists showed up for another working. Momma got to ride her own horses that day.

It wasn't Mother's Day proper, but it well could have been.

Moonlight Cowboying

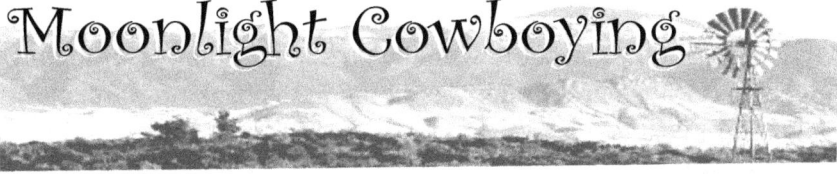

You don't have to own yearling cattle that get out on the highway only after you have gone to bed to appreciate the story I'm about to tell you.

What you will understand is first the humor, and then, just how often we ranch folk are thankful for so many little things.

Marci had been trying to fight off a record book-sized head cold for days so she wasn't in the best of humor and had only a little sleep, at best, for several nights.

About 9 p.m. on this particular night she took some cold medicine hoping it would help both the cold and the sleep problems. She tossed and turned, got up at 1 a.m. and took more medication and went back to bed an hour later.

At 2:30 a.m. the phone rang. That is never a good thing. It was their neighbor, Jim, telling her that they had lots of cattle on the Yankee highway and they were headed north up the canyon.

Marci slapped a still-sleeping husband, Frank, up side the head, mostly to wake him up but more so out of simple frustration. After 30-plus years of marriage she is sure he didn't know the difference.

Pulling on their cowboy clothes, they jumped in the pickup and drove up to the highway. Jim had gotten in front of most of them and had them headed back down the highway towards home.

The local sheriff was on the scene in his fancy car and was managing to hit Marci in the eyes with his high-powered spotlight whenever possible. When the sheriff wasn't blinding her, Frank was with his own Q-Beam. She noted that one-million candle power in your face at 3 a.m. is not soothing.

Marci was leading the cattle with the pickup and Jim was bringing up the rear of the herd. Frank was riding in the back of Marci's pickup for a fast get away. Marci was slightly amused that at this point, he trusted her driving. But then she realized he still hadn't figured out that she had slapped him awake.

They got the cattle to the gate of the pasture where they belonged, and as cattle will do, they changed their mind and headed back up the canyon again.

Marci wheeled out to go help Jim and both were trying to outrun the cattle up the highway. Frank was hanging out the back of the truck telling her something that sounded like "stop" so she hit the brakes. He had said stop but didn't intend for a slam-the-brakes kind of stop. He rolled over the side of the truck and recovered on his feet enough to block a side road off the highway.

The cattle finally went through the gate on the second try while the sheriff was still waving his spotlight around and trying to figure out who was on first.

They got home about 4:30 that morning. Marci's "Thank God" praises were for good neighbors, gentle cattle and a full moon.

How did Jim know the cattle were on the highway, you ask? He got up sometime after two in the morning to go to the bathroom and saw car headlights slowing down and weaving to miss the cattle.

Marci then thanked God for old men with weak bladders.

Know the Cowgirl by the Swish of Her Fringe

A cowboy or cowgirl's reputation is based on perception. It's a sad commentary to the world of high-top boots, but it's a fact.

It happened in the panhandle of Texas where it snows sideways and the top soil relocates daily with every norther that blows through, promising to freeze the backside off man and beast.

It is the land of wheat fields and millions of miles of electric hot-wire fencing.

Operators who lease wheat pasture for cattle are checker-boarded across the plains of Texas along with other misguided souls who want to run cattle in the worst weather this side of Fairbanks.

Misery and hard work make strong friendships among those foolish enough to think this is an honorable way to make a living. The wives are often fully half of the entire hired-hand crew and take at least their fair share of the head count to ride through and check daily.

Depending on who is telling the story, from his perspective, he is sure he has bigger circles to ride, sicker cattle to doctor, worse fences with longer drives to get it all done.

As happens in the small close community of 100,000 square miles, one wife will sometimes meet up with another wife in her daily ride through the cattle and decide to have lunch at the local Ptomaine Palace, also called the café.

When Billie Jo and Lynn walked into the café, everyone there knew them. Upon entering, Billie Jo would develop a slight limp and rub her shoulder. In casual conversation she would tell her stories that all began with "that Eddie Griego colt" or "that Hank Wiescamp condemned colt" – either of which had bucked her down and she was a little sore.

She always filled in the story with details of how he had "grabbed aholt of himself, swallowed his head, sunfished, jumped every which way" and finally got the best of her after a championship-style ride.

Nobody ever witnessed these rides, but nobody ever questioned those lash-batting blue eyes.

Lynn, on the other hand, regularly rode a Pasamonte Paul-Blondie's Dude horse, with justifiably reputed breeding for bucking. She admittedly was bucked off as soon as the horse pinned his ears back and as you might figure, there was always a witness.

"I saw you get bucked down the other day, but I came back by that place about an hour later and you were gone, so I figured you were all right," a café patron would tell her.

There were two reasons why people always took Billie Jo at her word. She had story telling down to an art and she always wore the correct costume.

Both women wore the same basic functional clothing, just with different style. Lynn's scarf to protect her neck from the cold was always cotton, large and wrinkled. Billie Jo's was bright-colored silk and always looked clean.

Both wore chinks but Lynn's were decorated with manure and orange marker chalk while Billie Jo's had silver conchos and long swishy fringe.

Lynn's spurs had regular length shanks with adequate rowels while Billie Jo's sported long shanks with pizza-cutter rowels that jingled when she walked.

Lynn always had a piggin' string in her belt in case she needed to tie up a gate or the muffler on the pickup to keep it from dragging. Billie Jo had two in her belt and two more bandolier style across her shoulder giving her a punchy look unequaled by any male counterpart.

Bottom line was that Billie Jo was regarded as a real deal, turkey feather in her hat, britches-in-her-boots puncher. Lynn was labeled as her husband's little helper.

Never underestimate the power of good story telling or dressing for success.

Ranching Sign Language is Universal

The rancher's wife stands at the gate waiting for him to make up his mind which direction he is going to go with the small herd of cattle he is bringing to the pens.

She sees him look at the cattle that are trotting a little faster than he'd like and then glance at her, but says nothing.

With a long established telepathy, she knows by watching him, she's got the wrong gate open even though it's the one he told her to have ready.

She slams her gate and runs as fast as boots, spurs and chaps will let her, to the other gate that is now the one they need open.

The language that is spoken, and more often not spoken, at the ranch requires visual skills as well as interpretive ones. Then there are some days the meaning comes through loud and clear.

Cattle and horses speak to their owners through patterns and nature's instincts. A mother cow will eventually give away the location of her hidden new baby if you just quietly watch her trying to not give it away.

She will look every direction but the right one until at one point she finally glances the way of her calf.

A baby calf, falling behind the herd while you driving them, will get a look in his eye that reads: in the next second you are going to see him with his tail curled up over his back, eyes glazed over, and leaving to go back to where he came from before you bothered him.

A horse's ears will perk up to attention while you ride through the brush and you can bet the bank he's heard, seen or smelled something you haven't. If the rider will pay attention, a horse will find more cattle in the brush than a rider will ever see alone.

Ranch husband and wife communications, while pretty much the same across the land, take on a bit more animation and sometimes humor. Well, it usually is funny later.

While she's chunking rocks at the bulls to get them through the gate and he's hollering it's the wrong gate, or wrong cattle or wrong something, the next rock chunking usually is directly at him. Not hard to interpret that.

An old rancher trick is to loudly give the wife instruction that she doesn't need, but that someone else within hearing does.

Rather than offend the neighboring help, he makes her look less than capable and hopes the one who needs to hear it, does. It usually fails in its intended mission and makes for a few days of SPAM as the main course at meals.

A nod, a whistle, a wave or a shake of his head speaks an entire language to his partner who most often is also the cook. Better judgment on his part is not always in use when communicating his thoughts.

You mean it didn't say, "for better or worse *and* mind reading?"

The Makeup Code

Not long ago I answered an SOS call from a female friend of mine who needed help pulling a calf. She had a first-calf heifer in sincere distress.

I arrived at the scene as fast as I could get there. I found my friend loading the necessary tools while giving me the details of the situation, as she knew it.

I was a little surprised to find her looking like she was going to town instead of to the pasture. I knew she'd been feeding cattle and checking heifers all morning so I was puzzled with the full makeup and nice hair-do.

And I, who had been in the house typing away at the keyboard, looked like a scene from the *Grapes of Wrath* and was barely presentable enough, even for ranch work.

I made no comment about her appearance at the time. I was just puzzled and felt perhaps she just didn't know better. You see, there is a code among ranch women who work out on the ranch along side what's-his-name.

Sometimes it is just easier to look more like ole what's-his-name than it is to maintain a daily semblance of a cover girl while eating dust, standing in cow manure and having your hair styled by 50 mph winds.

In my concern for this code-breaking friend, I mentioned the situation to another ranch woman friend of mine. She immediately suggested an intervention. She was worried that a coiffed, cute makeup-laden women in the pasture was going to give working ranch women a bad name.

She was even more concerned when she tried to reason the "why" of the presentation since this gal's husband had really poor eyesight. So what was the purpose?

The next time we, now named the Makeup Police, were to see this hard working but always-pretty ranch woman, she admitted to us she had showered, fixed her hair and put on makeup before she went to feed and check the heifers.

Before we could protest she explained it was so she could, afterward, make it to town in time for a meeting looking somewhat civilized, if only windblown.

In all fairness, it's difficult to not become hard looking doing the work that happens on a ranch. Efforts to hold some semblance of feminism are ongoing. When do you know you need to try a little harder?

A few years back I had gone to help ship some yearlings on a ranch where no one knew me except the people I had hauled with to the job. I was riding along trailing some cattle and one of the local hands rode up and started a conversation with me. The usual "howdy, my name is …" was first. Then came the curious, "You hired out to an outfit around here?"

I told him no, I was just helping for the day. But I knew if I looked rough, tough and punchy enough for him to think I made my living as a cowhand somewhere, I needed a trip to town and soon.

The Makeup Police have now recruited our latest suspect. She will forever be on the lookout for others who break code and show up to work cattle looking like they should be headed to a polite social event.

We have a reputation to uphold. How in the world will anyone at home know when we are going to town if we look nice *all* the time?

Things You Should Never Say to a Horseshoer

With the advance in rural living perpetuated by the invention of 40-acre ranchettes, trail ride associations and urban horse-owner playdays, the horseshoers of the world have found themselves in a completely new atmosphere of commerce.

Owning a horse is much like owning a thong bikini – anyone can own one but not everyone should. They are both something that should require some sort of an application process.

Farriers, or horseshoers as we regular rural people call them, have come from a long, dignified line of blacksmiths. Cowboys at the ranch usually shoe their own until they either are too old or they become financially sound enough to justify the cost of hiring it done.

Historically, a farrier was a horse doctor. It is only in the last hundred years that people who shod horses began calling themselves farriers and history is not clear on how that transformation came about.

It is unknown who invented the first horseshoe. Early Asian horsemen used horse booties made from leather and plants. During the first century, the Romans made leather and metal shoes called "hipposandals" and by the sixth and seventh centuries, European horsemen had begun nailing metal shoes to horses' hooves.

Around 1000 AD, cast bronze horseshoes with nail holes had became common in Europe. The 13th and 14th centuries brought the widespread manufacturing of iron horseshoes. Hot shoeing, the process of heating the horseshoe before shoeing the horse, became common in the 16th century. All of this before the first horseshoe was ever patented.

The first notable patent in the U.S. went to Henry Burden in 1835 for a horseshoe manufacturing machine. Burden's machine made up to sixty horseshoes per hour.

For those that are new to owning a horse and need the services of these hard working iron-pounders to keep your animal shod, here are some tips of etiquette or things you should never say to a horseshoer.

- Good Morning. Glad you are here. Can we reschedule? I have a lot going today.
- Can you bill me? I left my checkbook in the car.
- I know I said just a trim, but would you go ahead and shoe'em as well?
- I know it's been a long day. That's why I saved the worst one for last.
- I don't understand why the shoes didn't stay on. I had them done four months ago.
- Does it mean my horses have some sort of deficiency when they chew the paint off your truck like that?
- Oops, wrong horse.
- My weanling colt needs a trim. Maybe you could halter break him at the same time.
- I've got a new horse whose feet are in pretty bad shape. The previous owners said their farrier wouldn't work on him.
- I forgot you were coming. I just turned all the horses out.
- My last farrier couldn't finish. They gave me your name and number.
- If he didn't kick like that, I'd trim him myself.
- Can we shoe him in the arena? If he rears in the barn, he hits his head.
- Can you make it after 6 p.m. or on Sunday? I have to work.
- Good thing you are slow today or he'd have had shoes on when he kicked your truck.
- If you will just give each of the dogs a piece of hoof, they will get out from under the horse and quit fighting.

- ☐ Most times when he kicks, he misses.
- ☐ Can you shoe him so that he doesn't paw?
- ☐ If you get done in 30 minutes you'll be making $160 an hour.

Make every week "Be kind to your horseshoer" week. A good one is hard to find and harder yet to keep.

Two Women, a Cow and a Rope

Sometimes roping has nothing to do with skill with a rope. It can simply be fun, a challenge, or a survival tool.

When the neighbor calls and prefaces the conversation with "Feel free to say no," immediately after she says "Good morning," you know it is not a good sign for what is to follow. And even when you react immediately with a "No," you realize she didn't really mean what she said because she then tells you she already has the horses in the corral.

With anything resembling "man" power gone far away last weekend, just we "wimmen folk" were left here at the ranch. The project outlined by the phone call was to get to the corrals a big 4 year-old cow who had gone blind and not been to water for a few days.

For those of you that haven't tried it, cowboying a blind cow anywhere is not a feasible project. But we tried. She followed the fence the fair distance to the corrals and was within spittin' distance of the gate when she decided to go back from where she'd come and did so at a fairly rapid pace.

This is the place where it becomes not really about getting the cow in, but about not having to tell the men we tried and didn't get the job done.

That left the only option remaining to two of us who really didn't want to do it: rope her and put her in the stock trailer and haul her home. Short sentence, long project.

The roping was the easy part. Getting her in the trailer turned into a battle of wits, cuss words and strained body parts. The cow, while not wild, was not cooperative in the least. With no help whatsoever from her, we got her snubbed up close enough to have her head in the trailer and great hopes of finding a way to get the rest of her to follow.

Pulling her in the trailer with the traditional horse and rope method failed. It wasn't a pretty sight and unless you've done it, the explanation wouldn't make sense anyway. So with the horse tied elsewhere, we resorted to "brute" strength and sinister plotting.

We seemed to have an adequate amount of "know how" but fell short considerably with the physical ability to execute the plan.

The rope on the cow at one point had her a little short of air so she decided to lie down and quit completely. We loosened her up and after she revived she didn't want to get up on her feet. My partner suggested pouring a little water on her nose from her water bottle hoping it might make her feel better and she'd get up.

I doubted it would work, but what do I know. It could. So I poured water on her nose and hoped she'd either get up or drown and we could go home.

The water baptism didn't work, but a little nylon rope therapy did. She got up and it was shortly thereafter she got her front feet and most her body up into the trailer and we could see progress.

We were now totally committed. Everything but her back feet were in the trailer. We bodily tried to shove her in but no budging. Too much cow, not enough muscle from us.

I was getting tired and very aggravated over the cow's lack of cooperation. But with a little more nylon encouragement she decided that getting up in the trailer was better than where she was. It only took a minute.

I laughed when my friend said, "Well maybe that's why my husband just gets mad first." I agreed. A cow can make a man

angrier faster than just about anything in the world!

There is no glory for the rope burns I wear in places I can't show.
Three days of hot showers and ibuprofen and I'm as good as new
and with a new story to tell on myself.

And I have new wisdom. I either need to quit this line of work or
get in much better shape. I'm not sure which is harder to do.

Another Cowboy Job Well Done

The front wheels of the truck were almost off the ground and the hitch was close to dragging, but with cowboy confidence, the pair set off to haul this deal-of-a-century load of feed tubs to the backside of a remote ranch.

That was really when the trouble started but I'll back up and tell you how they got there.

Supplementary feeding of cattle in this part of the world is a necessary evil most years and is cost prohibitive enough to be a single source of going broke.

To add to the misery, the function requires dealing with a feed salesman. As a rule, they rank right up there with used-car salesmen and shady horse traders.

Minerals and supplements come in a variety of forms – licks, tubs, and blocks, to name a few. Donna had some cattle on a grass ranch on the east side of the state and her friend, Billy Jo was, this week, acting as a feed supplement rep.

After her pitch to Donna for the perfect supplement for the variety of cattle she had and the place she had them pastured, they made a deal and the tubs were to be delivered.

Billy Jo found herself a little short on equipment so she borrowed a former boyfriend's small, light pickup and small, light, and very old flatbed trailer to go with it. Refer to paragraph one and we are now into the story.

Billy Jo picked Donna up and they headed to the ranch. When they got to the gate, the steep canyon ahead of them was in view and Donna started to feel a little squeamish about the trip down to the bottom.

About half way down the cliff-hugging road, Billy Jo announced there were no working brakes in the pickup. About the same time, the trailer realized it too. It jackknifed, bouncing a good portion of the tubs on down the hill ahead of them.

Shrugging, they thought, "Well all right, we were going to take them there anyway." But the real problem showed up when they realized a portion of the tubs had landed underneath the truck.

These were forcing the truck off the ground, had the front of the trailer up in the air and the whole business stopped dead still.

The hitch was no longer workable and nothing could be dislodged. Cell phone service was not available in the remote location and nobody was expecting them for days.

And, of course, the jack was in the "other truck."

Using the only logic they could come up with, they decided the best thing to do was run over the tubs, hope the oil pan was in good shape, the brakes would suddenly come back to life and they'd go on down the hill.

In record time, they got to the bottom with no loss of life or limb. The truck hadn't been in that good of shape when they started out and Billie Jo had already broke up with that guy anyway.

The trailer had to be backed over a gully so that the hitch would come loose. It fell at an interesting angle and they left it there.

Everybody dusted their hands, knowing they had accomplished what they set out to do – deliver the tubs to Donna's cattle.

They limped the pickup back to town and Donna and Billie Jo parted friends.

Another cowboy job, well done.

Cowboys... You Gotta Love 'em

You Might Be a Rancher's Wife If...

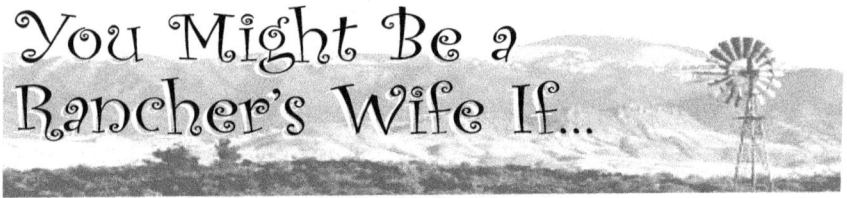

Sometimes a photo is worth a thousand words and sometimes it's living proof and validation of which I so often speak.

At the post office just today, I had to take such a photo. I just had to. Parked in front of me was the very thing you might envision of the rancher gone to town.

There sat a flat bed ranch pickup loaded to the gills. The back of the bed was heavy with a stack of mineral blocks. Squeezed to the front were a water jug, a cooler, an unidentifiable plastic box and two curious dogs who kept watching the post office door.

The headache rack was neatly organized with every thing a man might need while away from the house. A handyman jack locked to the frame, a couple of catch ropes not locked but tied up, chains, a come-a-long and assorted tie strings for tying gates shut, mufflers up or a calf down.

Now this version of a ranch pickup was newer than your average rusted beat up feed pickup that comes to town. He got his wife a good job in town and can now afford an upgrade – in pickups, not in a wife. You might be a rancher's wife if your job in town is considered a ranch subsidy.

Just a couple months ago this guy traded in his rendition of "beat up" for this new one.

56

And, as I understand it, for the first time in his life it's not a red one which is causing some mental anguish during the transitional period.

Meanwhile back at the ranch, you might be a rancher's wife if you know how to change the flat on your car but can't because the spare is on the flatbed.

It comes with the territory. You might be a rancher's wife if directions to your house include words like *miles, cattle guards, gravel road* and *last*. The word *last* can precede many things such as house, hill, left turn or cedar tree by a rock.

You might be a rancher's wife if your stock tank doubles as a swimming pool, the storage shed is a barn and you buy antiques because they match the rest of your furniture.

One thing a ranch wife's job in town works out for the ranch couple, besides subsidy, is the in-the-corral working relationship. Of which there is none. Wasn't much of one before she got the job but at least now the neighbors seven miles away can't hear the yelling and cussing. The dog finally quit hiding under the barn about a month after she went to work.

You might be a rancher's wife if duct tape is always on your list, the weekly paper comes a week later and the vet's number is on your phone's speed dial.

A rancher's wife will always have a shopping list that includes three sizes of filters, tires, chains, spark plugs and shotgun shells. And the best one, "get me a part that looks just like this one here," as he hands her wrapped in a shop rag, a greasy diesel smelling odd shaped thing he can't even name. "And make sure it's for the right year model."

Seeing that pickup today was like seeing it rain. It gave hope of better times ahead for ranching. As long as the wife can keep a good job in town.

So Ya Wanna be a Cowboy?

So ya wanna be a cowboy! In today's job market, being a cowboy isn't a very good bet. Ranches that need cowboys are dropping like flies to the pressures of government regulation and environmental restrictions. And, if that hasn't nailed them to the bank vault wall, high operating costs and a poor rain distribution plan has.

Still almost every little boy goes through a stage of cowboy dreams. At one time or another they've owned a stick horse, a set of plastic six shooters and a hat with a stampede string on it. They've pinned on a badge and chased the bad guys 'til sunset and roped wild cows 'til suppertime. That part of the romance of cowboying will remain alive as long as movies and television keep feeding it to the masses.

And those little boys? Many of them grew up to be big boys who still want to be cowboys. They've gotten college degrees, worked in big business, held down suit and tie jobs and managed a portfolio that may or may not let them afford to be cowboys.

The cowboy image is a marketing giant. If it's cowhide covered, cowboy picture adorned, silver trimmed, leather backed or framed in "barn" wood, it's labeled cowboy and it is surely almost pretty close to authentic.

The items are seen everywhere in every kind of store or catalog. But you can bet 99.9% of the marketers behind selling the cowboy dream, never stepped foot in any real cow stuff!

Back here on the ranch, where cowboying means the work we do with the cows in the spring and the fall, the romance isn't so fanciful. Between branding in the spring and shipping in the fall there are many months of that cowboy stuff that Hollywood hasn't discovered.

When was the last movie you saw where they were fixing windmills, digging up pipelines to find the leak and mechanicing on a pump jack engine? Hauling fuel to generators, driving a feed pickup, and welding back anything that broke last year are just some of those less than John Wayne kind of jobs.

And did I mention if you are working for cowboy wages you'll be poor enough to qualify for welfare, too? Not that any self-respecting cowboy would do that. Benefits? That would be the fact there is no time-clock so you are on call 24/7 and a brand new nylon catch rope is about as exciting as life will get.

Yes, you'll see first hand all those sunsets and sunrises the artists paint. And you'll see the 12 hours on either side of them on many days. Days off don't exist. There are just some days with less work to do than others.

I know I can continue to tell you all the things about cowboying you didn't read in the little book that comes in the Wrangler jeans pocket, but you aren't going to let me kill your dream. You still wanna be a cowboy.

And you know what? I don't blame you. It's the reason that those of us that are here are still here.

Pickup Inventory Time at the Ranch

January means inventory time to most places of business. Running late as usual, February was when I decided to do an inventory of the ranch pickup. This was to be inside the front of the pickup as well as in the back.

I also asked a couple friends of mine to inventory their pickups, just for a comparison. This will be a compilation of what we found.

In this part of the country, pipelines rule! They provide water where nature never intended water to be. So every ranch pickup will have an assortment of pvc fittings, pipe cleaner and pipe glue, and a dozen or so clamps of various sizes, usually all stuck in a little box with a couple rags.

Gloves, sometimes one of each right and left, but more often only one of several different pairs are stuffed under the seat, on the dash and in the glove box. Other under seat or stuffed items would be a pair of broken sunglass, piece of inner tube, a flattened roll of toilet paper too dirty to use, an empty binocular case and a piece of paper with a phone number on it, but no name.

Mail. There is always mail somewhere in there and usually an old envelope from someone you had already chewed out because you "never got it." It's not rare for the papers to be important things like the state land lease or the overdraft notice from the bank.

Usually you can find a gun or two in a ranch pickup and lots of ammunition lying around, but rarely the right ammo for the guns in the pickup. It's the same theory at work here as with the gloves. There will also be an assortment of book matches and empty Red Man sacks.

There are windmill leathers, preg-testing gloves, enough loose change to put a kid through college and a bottle of 10-40 oil. There is also a disgusting looking item that you have to use the channel locks that just happen to be on the dash to move whatever it is out of the way.

In the back of the pickup it's another world. Moldy hay beds the tool box that should have died years ago. It has holes rusted through it and keeps nothing inside dry or safe. The big well tools are slid up the side of the bed next to the propane tank that isn't used anymore but it gives the dog a place to ride and every now and then you can air a leaky trailer tire with it.

An old catch rope is coiled up and ready to use to tow something – either another vehicle or a cow. There is an assortment of chains, a headstall with no reins, a 5-gallon plastic bucket with some bolts and a little rusty water in it. The spare tire is anchored over the ball of the gooseneck hitch and cradles an assortment of pipeline parts in a plastic sack.

And of course, there are the universal items of survival – a big hammer, duct tape and baling wire. There is nothing that can't be fixed good enough for now with one or all of those three things.

Next it'll be the chore coat cleanout. It's spring and before it gets hung in the closet for the summer, it's prudent to check for live or semi-live items capable of decay.

Know the Cowboy by the Tip of His Hat

Cowboys, since the critters were invented, have been known for their good manners.

Tipping a hat to a lady, and according to the rule book, they are all ladies, using terms like ma'am and sir, and offering his chair to a woman or the elderly are all normal everyday actions for a cowboy.

There are two degrees of politeness with a cowboy hat. The hat-lifting or slight tipping of the hat comes when you greet a stranger. For friends and ladies it comes completely off.

Tipping your hat is done by barely lifting it off your head, by the crown of a soft hat or the brim of a stiff one.

A cowboy takes his hat off outdoors in several situations: when he is being introduced or saying good bye, as a greeting when passing someone he knows on the street, when talking with a woman, an older man or the clergy, when the National Anthem is being played or the flag is passing and at a burial in the presence of a funeral procession.

A man tips his hat in greeting to a lady, in acknowledgement to a lady when she thanks him, anytime he excuses himself for passing in front of her or jostling her in a crowded room, or when he asks a woman or elderly man for directions.

He always takes his hat off indoors except in public buildings or entrance halls and corridors of office buildings and hotels.

There are those that will say there is one place totally correct for a cowboy to wear his hat indoors and that is on the dance floor. That rule evolved within the last couple of decades when the onslaught of country music line dancers took to the dance floor. Real cowboys don't line dance.

Check out the old time dance spots and you'll find a "no hat on the dance floor" rule along with a place to put your cowboy hat for safe keeping.

Hat racks in public establishments seem to have gone the way of the horse and buggy, making it increasingly harder for a cowboy to part with his Stetson for the lack of a place to put it.

Laying it in the empty chair next you will expose it to a splash of coffee or a dollop of salad dressing from self or the waitress. On the floor under the table or chair, if there is room, opens it up to adornment with a foot print.

For those reasons, I've rested my criticism for cowboys eating a beef steak with their hats on. And the guy wearing the $500 hat? He's not the cowboy.

There is a long list of cowboy hat etiquettes that are just a way of life. I am fortunate to live in a part of the world where hats are still tipped and "Yes Ma'am" and "Yes Sir" are frequently used words.

Being polite means always being a little nicer than you have to be. One sign of good manners is being able to put up with bad ones.

And as Texas Bix Bender wrote in an etiquette book for cowboys, "Never go anywhere without your head in your hat."

Who Left the Gate Open?

It happens every now and then.

You walk out the door and you see an assortment of cattle in places they shouldn't be. Shouldn't be – in terms of that is not where you last put them.

At first, it takes you off guard. You blink repeatedly to make sure what you see is really those just-weaned steer calves that are munching the last of the green grass around the yard fence.

Why aren't they in the corral with the other 250 head eating good stuff to arrive at "steak" weight?

The next reaction, just before the panic, is to look toward the corral gate, only to see it standing wide open with nothing but air between you and all those calves who have not yet found the escape hatch.

Reality begins to take form quickly as you realize someone left the gate open. Not intentionally, just one of those brain-cramped-vapor lock moments when you think you did but you didn't.

It happens. Happens to the best of us and the not so best of us. We may even try to blame it on the "gate gremlins." Those are those unseen beings that lie in wait to open gates that are supposed to be closed and wreak havoc on the sanity and stability of a tired cowboy.

Every rural child in America is versed from birth on some basic rules in life. A very clear never-to-be questioned directive that is universal is "Always shut the gate." That mandate is echoed through the formative in years in a simple "Shut the gate, shut the gate, shut the gate."

I've yet to meet a cowman, cowboy or cow-wife who hasn't, at one time or another, made that inevitable mistake of leaving the gate open. Disaster doesn't always follow, but extra unnecessary work usually does.

If you happen to be the wife that left open the gate, the level of severity for the misdeed is heightened to what we would call today, Orange Alert.

First to follow the muffled cussing is a complete mathematical rundown of the cost per pound per minute per animal that is walking around the pasture instead of standing and gaining weight at the corn trough.

That is to be followed by the same calculations of financial loss when that animal gets out of a walk to escape being recaptured.

The statistical data can be never-ending and often is. With elephant-like memory the error is revisited on more than one

occasion for at least a year or until a bigger error occurs, whichever comes first.

Contrary to the beliefs established by the amount of discussion on the subject, the open gate is not always the fault of the Little Woman.

That lends itself to a completely different set of assessment criteria. Usually a shrug and an "I don't know what I was thinking but it's a nice day for a ride so we'll just put them back in the pens."

The who in "who left the gate open" is most definitely the determining factor for establishing the level of severity of the mishap.

The Cowboy Lesson in Fine Wine

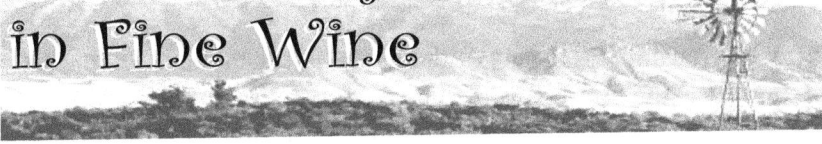

A pretty girl will stop a cowboy in his tracks every time. He will then do and say things he would have taken bets against if you had asked him prior to the pretty girl. This story is one of those times.

Ron was a good cowboy working for a good rancher with plenty of New Mexico country to tend. He was also aware this man had a daughter in an Ivy League college somewhere in the direction of "back East."

Winter had passed, heifers were about done calving, brandings were on the horizon and summer would soon be here. Life was good.

Then this cowboy's world turned upside down when Pretty Girl came home for Spring break.

The celebratory barbeque at the ranch gave Ron a little time to visit with Pretty Girl and he knew right-off she was way out of his social league as she chatted about opera, Broadway openings and formal dinner parties. The closest he could hope for was to hold his own at wine tasting. How hard could it be?

Like most cowboys, Ron liked to help the Colorado folks out with their brewery success and occasionally tried to help the Kentucky folks with their sour mash business. He knew he was going to have to get some schooling on the finer points of wine tasting.

Cowboys are experts at many things, capable of hard work with cattle, horses, fences, and equipment as well as making the hard business decisions required for a modern ranching operation. What they don't know, they aren't afraid to ask from someone who has a few more years and little more experience.

After conferring with a few of the hands in the bunkhouse that night, it was the consensus that an expert was required. Their collective thoughts pointed in the direction of the windmill man who was known to be able to fix anything and tell you a little bit just about everything.

In a phone call to this recommended universal expert, Ron was briefed on vintage, bouquet, body, sediment and all the various attributes of fine wine. The windmill man spoke with such knowledge and authority, the cowboy was duly impressed. He gave a brief pause of curious thought as to where this windmill man might have gotten his knowledge, but was in no position to question it.

It was clear his plan would be to invite Pretty Girl to share a little wine with him next time she was home.

Back to work he went, taking more notice than ever of the possibilities of the ranch. In his daydreams he envisioned Pretty Girl bringing him his supper after a hard day's work on the ranch he had married. By the time she actually came home again, he was in love.

It was summer and the cowboy invited her on a picnic to a pretty spot on the ranch with wine to be the main feature. They set a date and time and the cowboy whistled his way through his work for several days.

As will happen at a ranch, things didn't go as planned. He was down to choosing between a trip to town to get the wine or helping a late calving heifer through her ordeal. In a bind, he called the windmill man who agreed to bring him some wine in plenty of time for the big date.

Shined up, washed behind the ears and everything, Ron picked the girl up at the boss's house and headed down the road to the spot on the creek he liked best. They talked and laughed and the afternoon progressed about as smoothly as he could have hoped.

He might have actually realized his dream of capturing Pretty Girl and the ranch – if only the windmill man had thought to buy wine in a bottle instead of a box.

Who Left Murphy in Charge?

There is a school of thought that God is a cowboy. He certainly gets his share of "calls" during a cowboy's life and likely is the only one that has even a clue of how to get everybody out of the wrecks he gets invited to.

Now Murphy, on the other hand, is a full-time, large and in charge participant in every cowboy working. You all recall the dude called Murphy. He's the guy in charge of "everything that can go wrong will go wrong."

A cowboy won't ever have water problems when the weather is pleasant. The high heat of June seems to bring out the worst in pipelines and pumps while cattle lay shaded up at waterholes living for the next drink.

A January day when it is snowing sideways at 45 mph will almost guarantee a well that quits pumping.

General philosophy is if your water storage tank is full, the submersible pump will last 30 years.

Let the tank get below half and you may as well plan on calling the well man and hauling water until he can fit you into his schedule.

Cowboying is the only profession in the world where being "day help" is considered a career.

One must bear in mind that there is a reason day help does not have a steady job. When employing day help, one needs to remember that "day" does not mean the entire day to said help.

The fact that most "day help" won't work past dinner may account for the majority of the 3 p.m. lunches served at the ranch.

If the importance of the job to be done by the day help is red-flag high priority, you can count on him to show up riding a colt so green he can't turn him around in a four-section pasture and one that is guaranteed to buck through the middle of the shipping pen just before you weigh the cattle.

Murphy almost always stands in the gate, in the way.

Except, of course, if you need someone to block the gate to keep cattle from escaping, then Murphy has gone to the pickup to get another can of chewing tobacco.

At ranch headquarters, Murphy is a regular visitor.

The minute company shows up, the toilet quits flushing properly and the cattle at the corrals will break the float on the drinker, draining the storage tank so everybody, corrals and house, is out of water.

Try to do anything at the corrals in your boots and nightie, thinking you are the only soul within 50 miles, and everybody you haven't seen in six months will decide to drop by just as daylight sneaks over the hill.

Drive anywhere in that same get-up and you've guaranteed yourself mechanical vehicle failure and a long walk home.

Again, a road never traveled will find the neighbor moseying by, and of course, he'll kindly offer you a ride. Looking the neighbor in the eye the next time you meet is nearly impossible.

Headed to a meeting, a funeral or a command performance at the bank?

Murphy will assure you don't get off the ranch without first birthing a problem baby calf in the heifer pen or changing a flat tire.

And that, only after finding a jack that works, a lug wrench that fits and a spare tire that isn't flat, too.

It is at this point when you ask God to send Murphy to town. You're too tired to go.

Uh Huh, Sure He Did

I offer a caveat for the following story by saying "as it was told to me" simply because, while the source is quite reliable, the story itself is so wild, your first instinct will be to say "that's a lie."

Greg and Nancy headed out, stock trailer in tow, to get a neighbor's pink-eyed yearling out of their pasture.

They didn't have a real plan of any kind but they also didn't take a horse. The calf was so blind they figured they could sneak up on him and "coax" him into the trailer.

The neighbor the critter belonged to didn't know how to rope and Greg was still nursing his $27,000 and still counting... shoulder surgery. So Nancy was the designated roper.

Her plan was a simple one. Just rope the calf and let the rope go. No problem.

She eased up on him and surprisingly, even to her, caught him with the first loop. He was blind enough he didn't go very far; at least until the young overly-enthusiastic neighbor ran to pick up the rope and spooked the calf.

The blind calf, now wearing Nancy's rope and towing the neighbor, ran off with the rest of the cows to the other end of the pasture. Reaching warp speed rather quickly, the neighbor finally had to turn loose of the rope.

The calf, still on the run, made a big circle through the cows. Running and stumbling, he was more afraid of the rope than anything else. It was a monster he couldn't see but knew it was following him.

The calf appeared to be headed home to his proper pasture, but then he circled and headed back toward the cowboy crew standing at the trailer watching all this unfold.

Nancy made what at the time seemed like a smart-alecky comment, "Let's just open the trailer gate and maybe he'll load up on his own. He looks like he's heading right for it."

Still in joking mode, she moved to the end of the trailer and unlatched the trailer gate. The calf was still coming and at a pretty fast clip. She threw the gate back just in time for the calf to jump into the trailer.

They were all laughing very hard at that point. Nancy began claiming "Top Hand" honors when they realized someone probably ought to close the trailer gate.

That done, they were still in shock at the sight they had witnessed and were glad there were three of them to attest to it. Of course, then the discussion of where the credit was due began. Greg was sure he should have all the honors because he positioned the trailer just right on the road.

The neighbor claimed accolades for running the calf fast enough and far enough for him to circle back to the trailer and get in it with considerable momentum.

This exciting adventure took about half an hour and nobody had to unsaddle horses when they got home. It seems like if a day was going that well, they should have gone on to town and bought up some lottery tickets.

Telling that story to some poor West Texas winter wheat pasture puncher who is wearing an entire dry goods store on his back could elicit a violent reaction.

It's been my experience that any complaining done about the difficulty of loading sick cattle in a trailer never brought the highly unlikely moment of a critter loading by himself.

It did bring me a new trailer ball welded to the top rail of the trailer for me to dally a rope around for leverage.

Not everybody can be a "top hand." I'm glad I at least know a few.

Modern Medicine in the Cowboy World

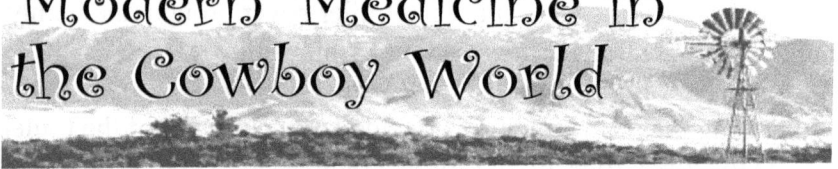

I told you no one would believe it, and many didn't.

In response to my story about the blind yearling calf loading up in the trailer on his own, one doubter wrote that he suspected the influence of Crown Royal or at the very minimum, an anesthesia overdose not-yet-worn-off the cowboy sporting the $27,000 shoulder surgery.

He called that the second lie. "Greg wouldn't spend $27,000 on shoulder surgery," he said. "He won't spend that on a truck. If you don't believe me, ask him about "old red.""

I responded by explaining to him that I trusted his assessment of his close friends but that the Crown Royal was very likely only available for medicinal purposes for those with refined taste preferring it over pain meds.

I also explained to this occasional ranch visitor that cowboys are sometimes the biggest babies – too tough to take the doc's advice or medication, but world class at moaning and groaning for the 90-mile-drive back to the ranch. It's not unusual for the Mrs. to grab the pain pill bottle saying, "Give me those blasted pills! One of us needs to feel better."

As for the $27,000 shoulder, most cowboys will sell their soul to get a body part fixed so they can go back out and do whatever it was they did to hurt it in the first place.

Another cowboy, on the wise-side of his fifth decade, had a stout three-year-old colt buck him off, resulting in an emergency room visit. This was followed by time spent with triage nurses, doctors, radiology technicians, family practice physicians, orthopedic specialists and a bona fide physical therapist.

His wife carried a dictionary around to translate their diagnosis, prognosis, treatment protocols, medication and device advice. This was followed by a barrage of bills in the mailbox that took a fair amount of accounting expertise to decipher.

The real problem at hand was getting to the cure. His actual diagnosis was Type 2 acromioclavicular separation, as in "hurt shoulder." That made logical sense as that is where he landed. If he had just had the foresight to find a soft spot to land all this could have, in theory, been avoided.

Each of the specialists, with a serious direct eye-to-eye gaze, told him to wear the immobilization device. We call that a splint. They advised he not lift anything including his arm and it would be six weeks before he could move anything except his lips to moan.

Next came the electric stimulation to the muscles to facilitate healing and a very dedicated physical therapist determined to bring healing no matter the pain level. In a moment's time, the cowboy was promoted from complete immobility to lifting weights over his head.

A series of repetitive moves with pulleys, weights and other devices ensued, moving the cowboy into a realm of exercises he couldn't have done before the accident, let alone while on injured reserve.

The cowboy declared there was nothing about roping that was as physically hard as what the therapist had him doing. So he went home from therapy, saddled his good horse and roped a pen of steers just because he could.

Hee Haw's multi-talented Archie Campbell played many roles on the 60s–70s variety program, one of which was the leering doctor giving sage advice to his patients. "If it hurts when you do that, don't do that."

The jist of all the medical advice given to the cowboy is exactly what Dr. Archie was saying. If it hurts, don't do it. If the cowboy had just remembered *Hee Haw*, he could have saved a lot of money.

Holding Herd is Not Administrative Work

Fall is sorting time in ranch country. The mechanics of the job vary from ranch to ranch but with the same end goal in mind: start next year by keeping only the best of the herd.

When the sorting is done out in the pasture as opposed to in the corrals with a proper sorting alley, it can be a little more "western."

The cattle are driven into a corner or thrown up against a fence somewhere in the pasture and surrounded by the cowboy crew. One or two of the executive class cowboys do the cutting (sorting) horseback.

Holding herd for a cattle-sort of any kind is often considered menial labor. I suppose if you take into account you sit for hours using not much brain power, enduring the dust, wind, heat, cold and butt-chewings from those cowboy executives, it is pretty menial.

The dignitary cutters get together in a little meeting in the middle of the herd and the conversation goes something like this:

"What do you think about that one-horned cow over there?"

"She's pretty poor and more than a little long in the tooth."

"Calf is probably stealing milk from the other cows to live."

"Don't you reckon we can get one more calf out of her?"

"Probably. A guy sure don't want to ship any of his good cows."

Time tends to drag when you are holding herd. You'll see the pocket knives come out as herd holders begin to carve on the calluses on their hand or clean their finger nails like there will be a hygiene inspection later.

The tobacco can lids flash in the sun as chews are freshened. The athletics of spitting in every direction including between your horses ears is a skill worthy of an Olympic event.

Every now and then, one cowboy will ease over to another and strike up a conversation all the while keeping one eyeball on the herd so as not to be slack in his duty.

Harassing the cutters is another way to pass the time. It is often passed off as attempts at being "helpful." When cutting dry cows out of a herd of black cows you'll hear that helpful cowboy holler and point in the general direction of 250 cows.

 "Hey, that black one is dry, the one with the black face and two ears."

Then there is the cowboy with the incurable roping disease. He lives for the moment something, preferably something with a nice set of horns, eases to the edge of the herd and wears any appearance of an "escaper."

He must create the impression he's doing all he can to hold that animal where he belongs while in fact he is doing his best to drive him off. Once the critter has made the break, the cowboy unlatches his rope string and builds himself a loop while putting the spurs to his horse and leaving in hot pursuit.

At some point later this same cowboy will tell the boss with only the sincerest look on his face, "I sure hate to rope your cattle, but that yearling just had it in his mind to quit the bunch."

The boss lets him off the hook with "Oh that's okay, those things happen. I didn't mind the first time or the second or even the third, but that last time that steer darn near got back in the herd before you got your loop on him."

Busted by the boss, the cowboy will pull his hat down, shuffle his feet and mutter something about maybe needing to get a faster horse.

Angels in Cowboy Boots

The year 2007 started with the plains of Eastern New Mexico and Colorado buried in several feet of snow and 10-12 foot drifts separating ranchers from their livestock.

Hay was dropped from helicopters to stranded cattle and the event made national news. Like a fingernail across the blackboard, eastern newscasters slur western pronunciation of words like La Junta or Hereford while cameras roll documenting the blizzard.

It is times like these that stories emerge telling of miracles that can't be explained and people helping people with a kindness most of the world has never seen.

One ranch family in eastern New Mexico opened the doors of their tiny house and it became home to 44 people stranding during the blizzard.

Divine providence provided that one of those guests was driving a truck loaded with foodstuffs, providing basic sustenance for all for several days.

Every now and then, something happens to all of us that gives us cause to know we aren't alone in this world even when it appears that way.

Some years back, out in the middle of nowhere New Mexico, in the middle of the night, two cowboys were headed home from one ranch to another to the south. It was past 11 p.m., on a dead-cold winter night.

They had been lion hunting all day, but as the norm, dark wasn't a deadline for the hunting dogs. The pair had to wait late into the night for the hounds to return to the pickup so they could head home.

Snow had fallen the day before and a good amount of it was still on the ground when they left the other ranch headquarters and headed south. About half-way into the 14-mile drive, one of the cowboys was beckoned by the call of nature. Having just left one ranch headquarters and so near another, it didn't seem like it should be urgent but apparently it was. They stopped and stepped out into the cold starlit night to "visit" the backside of the horse trailer.

Out of the darkness, back a distance from the road and from beneath a big cedar tree, came a weak voice repeatedly calling "Help! Help!" Both men looked at each other, not sure if they were hearing things or if perhaps one of them was losing their mind.

One cowboy with a tad of streetwise instinct was sure it was an ambush and a set up. He reached into the pickup for his pistol. The other cowboy wandered toward the voice, without pause, wondering, "Who in the world would be out here in the middle of the night?"

Frail, thin and 80-something-years old, John had taken a drive through the backcountry roads, headed nowhere, but as he later told them, he did this often just to see the country. That afternoon he had gotten his car soundly stuck in the mud and snow on one of those remote dirt roads. He wasn't sure where he was but soon decided to attempt to walk somewhere for help. He had no idea where the nearest ranch house was or which direction he should go.

It was a warm afternoon when he started out so he didn't take a coat with him. He wore only a light shirt, cardigan sweater, dress slacks and dress shoes with thin socks. His footwear got wet early in the afternoon and as sun began to set and temperatures dropped, a chill settled over him. But John kept walking.

The thermometer fell into the teens that night. After walking for hours in the dark, cold, miserable and totally disoriented, he said he curled up in a tree and prayed. He later told his rescuers he believed he would die that night and be found huddled in the cedar branches.

Sometime later, he heard the cowboys' pickup and trailer as it rattled down the road. Summoning what little strength he had left, he again began calling for help, knowing his prayers had been answered.

In the middle of the night, in the middle of nowhere, and from someone that, under normal circumstances, would never have been there at that place at that time, he was rescued.

No one involved that night thought anything that had happened was a coincidence. The cowboys, unable to vocalize their thoughts, shrugged it off as best they could.

But John never quit giving praise for his answered prayers. He knew his prayers had summoned two angels to save him – and this time, those angels wore cowboy boots.

Bein' Neighborly at the Mud Pump Bar

Some words are just fightin' words and there is no way around it. This is about one of those times.

Four West Texas cowpunchers had spent the week at the end of nowhere gathering and branding cattle. They were tired, cold and running on empty by the time they headed for home.

The eagle had flown that day and with a few spare coins in their pockets they were feeling a little flush. While not believing in government aid, they did believe in being neighborly. It's the cowboy way.

Their justification was this.

Colorado joined Texas making them neighbors. For a good while these four had been neighborly underwriting that pure Rocky Mountain spring water brewery. In fact, another wagonload had just been sent down the hill to Texas and these cowboys were feeling dutiful to their cause.

The first place they came to looked like it would fit the bill and was in fact the only place between the prairie dog town fork of the Red River and home. It was called the Mud Pump Bar.

Not being totally unobservant, they had seen a little of the oil field activity in the area. Comment was limited to "those pump jacks make a good place for the cattle to shade up in the summer."

That there might be "oilies" coming with this oil field activity had not yet crossed their consciousness.

The punchers parked their horse trailers out of the wind as best they could and proceeded into this fine establishment in an orderly fashion.

Now the Mud Pump Bar was a high class joint. Just last week they had changed out the sawdust on the floor and there were several new egg crates on the ceiling. There was a jukebox with country music, a waitress in a short skirt and so many beer company clocks you could tell time from Amarillo to Corpus Christi.

When their eyes adjusted to the low ambient light, the problem they didn't know existed became quite visible. There was a long table in the middle of the room filled with oilies.

Being basically friendly and very thirsty, the foursome decided to ignore this breech of bar hospitality.

Sitting alone at the bar was one old man dressed in Wranglers, boots and a cowboy hat. He struck up a conversation by noting the four newcomers were also cowboys. "Hard life, ain't it," he said in more of statement than a question.

The cowboys were about to agree but before they could launch into a hard life story, one of the oiles jumped up to challenge them. This guy, evidently for most the day, had been irrigating an already foolish mind with a sudsy beverage.

"Man you don't know nuthin' about hard work. You ought to try following us around one day."

The table of oilies got quiet and for all appearances, to a man, they intended to back their mouthy buddy.

While seriously outnumbered, the punchers weren't ready to have some oilie telling them about hard work. But they let it ride this time thinking the moment would pass.

Bent on self-destruction the oilie made another run at the cowboys. "So where'd you boys park your little ponies?"

In Texas there are only men and horses. While a cowboy's hide might shed insults to himself, nobody made fun of their horses.

After the sheriff left and the oilies had paid for the damages to the premises, the four cowboys went on about their business of helping out those Colorado folks.

 It's a dirty job but someone had to do it – they were just bein' neighborly, you know.

When the Cowboy Was Sent to China

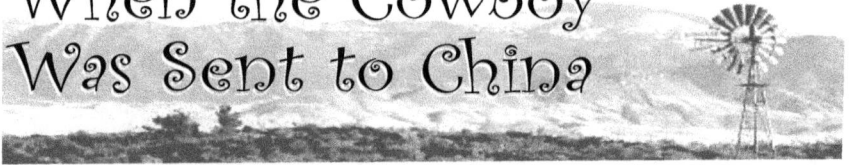

Everybody in the world knows that cowboys settled, civilized, and developed the West as we know it today.

Right after the dust settled in Dodge City, they went about their business of planning out vast cattle ranches, the petroleum industry, the space headquarters, top ten ranking U.S. cities that are still called "cowtowns," invented rodeo, and, of course, were the draw to gather the prettiest girls in the world.

They gave clever names to every inhabited spot on the trail never knowing how handy that name might be to future generations.

Our cowboy in this story was young and single and, while having a reputation of sharing his talents with various girls in the area, had fallen into a routine with one particularly pretty one. He worked by day for a big diversified cattle outfit whose operation ranged from buying, back grounding and trucking, to pasture cattle and feedlots.

The cowboy and this little-pretty would go together to dances, calf fries and an assortment of ropings. He was fine with the arrangement and, besides, his job left him limited time to shop around. He didn't give it much thought, other than from time to time, he would notice a dangerous gleam in her eye.

Ignoring the warning signs, he rocked along until one day he came home to find she had actually cleaned his house. His saddle no longer resided on the living room floor. Gone were his extra headstalls, bits, boots, leggings and spurs. She actually had them hidden in a closet!

The clincher to the deal and cause for panic were all the candles and little dishes of potpourri she scattered around here and there. No self-respecting cowboy could allow such things. The next day he took the boss aside and explained the situation. The boss, being male and having been single at one time, understood the immediate need and supplied the solution.

Being basically a gentleman and blessing his ancestors for their foresight in giving names to towns like Palestine, Iran, Germany, etc., the cowboy told the girl that his job was going to require him to be sent to China for an undetermined length of time.

This poor geographically-challenged girl was not aware that the China currently referred to was located on the Louisiana border of Texas. Nor was she aware that the undetermined length of time was between three and five days. She evaluated the situation from her perspective and decided not to wait for him to get back.

The cowboy denies that the girl was the reason he was sent to China but he did go thinking it was a big company promotion. He had visions of rolling into town in one of the big company cars, carrying a big company checkbook, giving the hired help directions while sitting down to write the check and then coming on home.

What really happened was his boss told him to get his backside in one of the cattle trucks headed that way. He utilized the sleeper in the truck for most the trip and they rolled into China just before daylight.

The man in charge of the cattle operation in China met the cowboy at the truck and said, "You see that gray horse over there?"

"Yes sir," said the sleepy cowboy.

"See that saddle in the tree?" said the cowman.

"Yes sir," parroted the cowboy.

"Well get them together and get your butt on them."

The cowboy was the only help for a four-day gather and pairing-up deal as well loading them on the trucks to ship out of there. At night the man, along with his three pretty daughters, would sit around playing IQ games. The cowboy lost out on both counts.

I'm sure there is a moral to this story but it would do no good to figure it out. There will always be another pretty potpourri-toting girl and therefore another cowboy looking for a job in China.

Cowboy Manners

We live in a time when rules and regulations are everywhere. If the government hasn't regulated it, we the people have. We write rules and we pass out manuals with job descriptions and the laws to abide by while on the job.

The job a cowboy has is not just a job. It's a heritage that has evolved over centuries of man working with beast. The cowboy evolution became about man, horse and cows. It also became about the type of man who was, and is, a cowboy. I'm talking about the cowboy named after the cow, not the cowboy named after the event at a rodeo.

The man that came after the invention of the cow came with a set of rules, a code of sort, that hasn't been written up in books. They aren't in a manual handed out at the bunkhouse. But they have been passed from generation to generation among the cow hands and between father and son.

These are laws of respect and etiquette that weren't taught, they were just lived. Those laws still hold true today, although the awareness of them seems to be fading as more and more cowboys are "found," not raised.

There have been some real genuine legitimate indisputable cowboys in my life. With my tendency to write things down on paper, I interviewed them over the years and asked them to tell me what it was I should know in case I'd missed what it was they thought I knew.

Here are a few of the etiquettes they all agreed every cowboy should know and do, no matter what.

1. Never ride another cowboy's horse, unless it's a matter of life or death.
2. Never use another cowboy's equipment without permission.
3. Never ride between another cowboy and the herd. Always ride behind them to get where you are going.
4. Don't ride in front of the boss. He knows what he wants to do and you don't. If he's tracking cattle, stay back or you'll mess up the tracks.
5. Never ride into a herd without being asked. If you are holding herd, hold the herd…Period.
6. Don't ask the boss what you are going to do the next day. If he wants you to know, he'll tell you.
7. Always take care of your horse before you take care of yourself.
8. Always be on time. Nothing makes a cow boss quite as mad as having to wait on someone.
9. Cowboy, take that hat off! If you are in the presence of a lady or if you go into someone's house, show your respect and hold that hat in your hand. Watch your language in mixed company. If you are sitting in a room and a lady enters, stand up.
10. Always help the cook with wood and water and don't ever get into his grub unless he asks. Always put your plate and silverware in the roundup pan (dishpan) after you eat.
11. Don't ever take a dog to help another outfit. They may not like dogs. Never yell at another man's dog.
12. Always roll your bedroll when you first get out of it. ALWAYS leave a clean camp.

The best advice my mentors could offer was to always be respectful, dependable and do your best. Manners count for cowboys too. There is no one finer to be around than a gentleman cowboy.

Cowboys and Technology Struggle to Partner Up

An oxymoron is when words that are contradictory and incongruous in nature are put together. Some examples of those are *deafening silence, government intelligence* and *cowboy technology.*

We live in a world where a cowboy's job, in many ways, is as basic as the day cows were invented. Just as true, 21st century technology is being forced on them one database at a time.

Fifty years ago, cowboys were still trailing cattle to the rail yards to ship them to market. The basics of cowboying then were the same basics as 50 years before that. You needed at least one good horse and a few knot heads, water and grass for the cows and a dry spot to unroll your bedroll.

Today there are still cows and there are still a few good horses and a quite a few knot heads. There are still a few ranches where cowboys "go out the wagon" and sleep in a bedroll. But for the most part, the simplicity of those days is gone.

The rancher, cowman and cowboy are usually one and the same these days. Technology in the cow business is no longer just what those "boys at the college" do. It is teetering on the brink of requiring those sweat-soaked-Stetson-wearing icons of the plains to carry Palm pilots and check their email.

Feed and fuel prices are high and a decade of drought has ranchers continually assessing feed programs. As the cow business is fine-tuned tighter and tighter, ranchers are relying on "data" to make decisions that before required only a "good eye."

The Country of Origin Labeling (COOL) is looming on the horizon and cattlemen will be required to provide electronic tracking of their cattle from pay day at the ranch scales to the meat counter at the super market.

This creates some humorous mental images. There are some hard heads out there that run quite a few cattle that are still struggling with the technology of remote controls, VCRs, fax machines and cell phones.

The very idea of them having to keep track of and keep in working order any kind of hand-held computerized device is science fiction at its best.

There are countless pocket calculators, tally books, pens, cans of chewing tobacco and eye glasses laying in the bottom of water storage tanks and drinking tubs across the West.

It only takes one broken float or one plugged standpipe for that stretch-and-lean over the tank to make repairs that will sacrifice those shirt-pocket treasures to the deep.

And what if he carried it in the pickup for safe keeping? It could work if the oil doesn't spill on it, the dog doesn't chew on it, the dirt doesn't bury it and the hired hand doesn't kick it out on the ground as he bails out to open a gate.

There doesn't seem to be any relief in sight for these digitally-challenged cowboys. Tight management practices and continued scientific data processing in the cow business will eventually push the majority of them into the cyber age.

The partnering of cowboys and technology is a marriage that gives new meaning to the term "shotgun wedding."

Cow-Calf or Yearling Operator... How to Tell the Difference

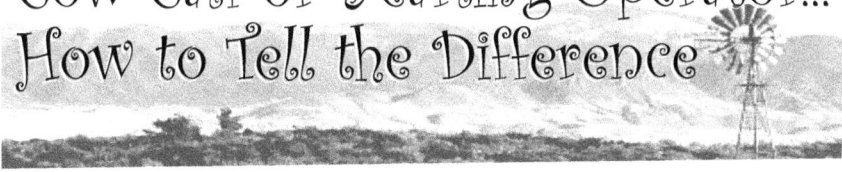

I drove into the yard last Sunday evening and heard a serenade that can only be called "The Song of Money."

The calves that had been weaned that day were standing in pens bawling their little fool heads off just sure their lives had taken a turn for the worst. Their mamas were not far away singing the same song but in a lower, louder voice.

To the average person not accustomed to the cattle business, the ongoing cries that continue round the clock for 3-4 days would perhaps be annoying and melancholy. It is music to the cow-calf man's ears. It is the sound of a nearby pay day.

Weaning time is part of the business most familiar to the cow-calf man. Yearling operators don't have to mess with the emotional details of telling an overgrown brute of a calf to kiss it's mama good bye, that's he's taking a ride on a truck.

Most people outside the cattle business don't know the difference between a yearling and a cow and are perfectly content to leave it that way.

Neither are they aware of any differences between several classes of cattlemen, such as the cow-calf operator and the yearling man. In some cases the cattleman will wear both hats.

At a distance the two look pretty much the same. They wear boots, listen to the same radio station and hold similar opinions on politics.

Close inspection of the back of their pickup is often a good place to pick up a clue. A cow-calf operator will have some empty feed sacks, a pipeline repair box of stuff, a beat-up tool box, some baling twine, a shovel and an axe if it's ice season.

A yearling operator will have electric fence posts, chargers, batteries, crumbs of alfalfa, and three hundred pounds of baling wire.

The guy you run into at the post office with red eyes, a runny nose and who speaks in a hoarse whisper, is probably a yearling operator. He has the same shipping fever he's been doctoring in his yearlings for six days and he looks worse than the cattle.

Close inspection of his clothes is another sleuthing technique. A cow-calf man will have cottonseed meal dust in every fold of his clothes. The yearling man will have speckles of blue ink all over him.

It's not ink but pinkeye medicine. It is a given that more purple violet color will end up on the man's pants, shirt and noose of his rope than ever gets in the eye of the animal.

In any gathering of cattlemen you can look closely at their hands and forearms and pick out the yearling operator. The guy with all the tattoos up his left arm, yep that's him. But those aren't actually tattoos – they are ink pen tally marks.

His Monday count is close to the palm of his hand, Tuesday's is on his wrist and by Thursday he's nearly to his elbow. Course that means he's only washing one hand before a meal. That's another clue.

All winter the cow-calf man rides around in the feed pickup with all he needs in the cab next to the heater. The yearling man is out horseback in harsh elements with two ropes, three pigging strings and saddle bags full of medicine.

Like the mailman, neither wind nor rain nor dark of night will stay this cowboy from the swift completion of his appointed rounds.

He'll be the Popsicle in the Carhart coat frozen to his saddle with icicles hanging off his mustache long enough to freeze to the zipper in his coat.

It is simply some more of that glamorous cowboy life that didn't make the brochure.

Don't Drink the Blue Stuff

In this season of winter I find myself giving some thought to things I am thankful for while the wind howls and thermometer drops.

At the top of my thankful list is how thankful I am that there are no heifers to calve to this winter.

Making a mental note of the advantages of calving heifers when nights hit the single digits on the thermometer makes for a mighty short list. Frozen hands covered in afterbirth goo and wallowing a wet calf to the barn with a mad momma intent on snorting you out of the pen seem to not make the list.

Birthing anything, calves or children, never seems to come at a convenient time. I have no end of recall to situations where the birth of a child aggravatingly interfered with hunting season, an important roping or happy hour at the local cantina.

Why even as recent as last week, a local basketball coach had to miss the first game of a tournament because his wife thought she needed to have the baby at the hospital. She turned down the offer of the school's hospitality room, even though she was assured of a cleared table and at least an hour of privacy.

Having given birth myself has given me the sense that I have one notch more empathy for a heifer birthing a calf than any male counterpart could ever muster.

When those calving chains are hooked to a come-a-long to pull that unwilling calf on out, I give audible praise to God that the cowboy standing next to me hadn't delivered his own child.

Cowboys, for all that they are, there is a list of things they are not. Two of the "nots" are empathetic and patient when it comes to the details of womanhood.

Mentioning that the cold is making your skin dry and flaky will return a cynical comment about the hot oil treatment you had only a week ago. And they truly think getting showered in hot hydraulic fluid when the line broke on the tractor while you were in close proximity should have fixed that dry skin problem.

In today's world many ranch wives have town jobs and that sometimes leaves the cowboy boot clad Rambo of the outfit in charge of feeding, dressing and monitoring small children.

"What do you mean tortillas and M&M's aren't a complete meal? Besides, today we finished off the bag of those cookies with all natural ingredients."

You will never convince him that 12 hours of the Cartoon Network is damaging to their minds. His alternative is the fishing channel or *Baywatch* for its water safety tips.

Revenge is often unintended but just as sweet. Let's say the cowboy of the house is a Kool-Aid drinker along with the kids. The gallon pitcher on the counter invites him to pour a big glass full and down it quickly to sooth his quenched parched thirst.

When he finds out that blue stuff was not a blueberry punch but in fact a newly mixed batch of Miracle Gro intended for the house plants, all hell breaks loose. By his decree, any beverage with a blue tint is forever forbidden to exist on the ranch.

The real downside to the event for the cowboy is he will have to endure years of ribbing from his buddies on just what could possibly happen after drinking a full 12-ounces of Miracle Gro.

The Curse of the Gal-Leg Spurs

It was August in DeLeon, Texas. The annual Peach and Melon Festival Parade was about to begin.

Following the usual train of thought for the state of the union mentioned, this was the best darn parade to precede the best darn peach and melon festival in the best darn part of Central Texas.

Garland and his brother, Jerry, were sitting on their horses watching the parade and waiting for their turn to get in line at the end. The cowboys had come to town to be in a parade. No one ever really knew why.

Jerry was riding his good roping horse and Garland was on a fairly green colt. For you pilgrims, green wasn't his color but his level of experience.

Garland owned a pair of gal-leg spurs he was right proud of and wanted to wear in honor of the occasion. Gal-leg spurs are called such because the shank on the spur, the part that sticks out from the boot to hold the rowel, is shaped and engraved to mimic a saloon girl's leg. The design dates back to early spur makers in the late 1800s.

While Garland's spurs had never been on a boot prior to parade day and they wouldn't stay in place, he wanted to share this piece of history with the pilgrims of DeLeon. Occasionally he'd have to reach down and put his spurs back into proper position on his boot heel.

The parade was pretty long and the colt Garland was riding got progressively more nervous. Once in awhile Garland would walk him off and then ride him back to the waiting spot, trying to settle him down.

After a time of watching the parade, the colt's nerves were knotted up tight. Garland reached down to adjust his gal-leg spur and the colt came completely untrained. He crow hopped (an unserious form of bucking) around in a circle.

Ever vigilant, Jerry reached over as the colt bucked by him and grabbed the cheek strap of the colt's bridle. Garland had landed somewhere behind the colt, somewhat ironed out on the pavement.

A lady from the crowd, not realizing Jerry had the colt caught, came running toward him in an attempt to help. As she got closer and closer, the colt started backing up without actually moving his feet. The colt knew Garland was behind him so his attempt to avoid the lady turned into a stretched out squat.

Finally, as the lady got right up close, the colt just sat right down on Garland's chest. Nearby paramedics were there in a flash and began an immediate assessment of the situation. They asked Garland if he had any pain.

A strained reply from Garland was, "You damn right and I know exactly what this horse weighs. Get him off me!"

The area was cleared and medics ascertained that Garland had a slow heart rate. A lady opened his shirt and started rubbing him down with ice.

The paramedics told Garland that this was a public function and by law he was required to go to the hospital as a precautionary measure. The seven-block ride to the hospital would cost $580.

Jerry asked Garland if he was okay and Garland said he was although he knew he'd be a little sore tomorrow.

"Well then," said Jerry, "get the hell up and let's go home."

They were not invited back the next year to ride in the DeLeon Peach and Melon Festival parade.

Famous Next-To-Last Words for the Cowboy

The mind-set of never turning down a rain when you ranch in the Southwest has been pushed to the limits this year as ranchers saw almost double their annual average rainfall arrive all at once in a month's time.

Slow, if never, to grumble, ranchers have fixed water gaps that have been solidly in place since the last millennium, repaired washed-out roads repeatedly, and found leaks in the roofs of homes, barns and outbuildings that didn't exist until, of course, it rained.

And still the rains came. Most recently, the moisture that was so welcome this year was falling on the backs of newly-weaned calves with the threat of bringing on job security to the guy in charge of doctoring sick ones and hauling off dead ones.

The tens of hundreds of cattle trucks scheduled for dirt road destinations will be standing by waiting to see if it is really going to happen the morning after Mother Nature has again dumped inches of rain on ranches that now have more grass than anyone has seen in their lifetime.

Making fall gathering and other assorted seasonal cattle work cold, miserable and hard to plan, the misery is simply accepted as part of the business. No one in the business dares wish it would stop raining. Who would accept the responsibility for such a bold statement if indeed it did stop raining for too long, again?

You can count on a few things as a cowboy and usually they have to do with those types of off-the-cuff statements followed by results that become legendary.

- "Weatherman says there is a storm coming today but we'll be finished long before it gets here." Result: The hundred year blizzard hits just as the cowboy crew arrives at the backside of the ranch and 20 miles from headquarters.
- "Send the city kid that came to help to the north end. No cattle ever go up there." Result: One pilgrim struggling with the entire bunch of cows and calves while the real hands hunt for cattle.
- "It'll be a good day to ride the colt. We don't have any serious cowboying to do." Result: Hunters left the gates open on four pastures and 600 head of cattle are mixed or missing.
- "Don't worry, they can't get plumb away. There's an ocean on both sides." Result: Four hard days of looking for wild cattle in heavy brush-covered country.
- "Don't worry, they won't get away. They're afoot and we're horseback." Result: A corral of ridden-down horses, tired riders and cattle still running wild and free.
- "The break-even on these cattle won't pencil out right now, but the market is bound to improve before shipping time." Result: The bank says they will extend the operating note one more time.
- "That colt never saw a day he could buck me off." Result: The wife getting quite handy at doing all the riding, doctoring and feeding while he heals-up.

Another list for another day is the really incredibly "un-wise" things a cowboy will say, without thinking of course, that will land him in the dog-house and eating bologna sandwiches for an undetermined amount of time.

That list belongs in the "last cowboy words" category and usually starts with some brilliance like "What that woman doesn't know won't hurt her...."

Rodeo and Ropings

Creative Cowboy Math for Ropers

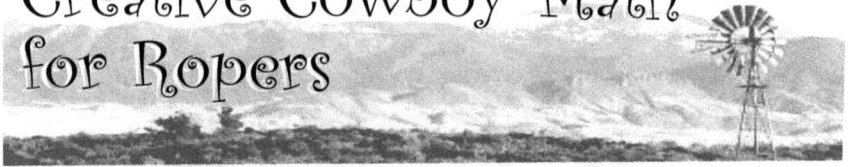

Every now and then someone forces the cowboy to do a little scratchin' on paper. He'll call it a tally sheet and prefers to keep his figuring on his left hand glove or the right leg of his chaps.

For the purpose of this story, I'll target the team ropers. Don't worry; they are used to the abuse.

If someone with a bookkeeping background were to put the ropers "figgers" on paper, it would read something like "Income and Expense Statement, Profit Center: Competition Roping."

The expense column would have a long list of "must haves" that total to a shocking number. The cowboy will qualify the sum with "estimate only – exact records are not important."

It is hard to tell which comes first, the rope, the horse, or the rig. They are listed here in no particular order of importance.

Expense:

Top notch #1 winning rope horse	$10,000
Back-up practice horse	$9,500
Three-horse slant aluminum trailer	$30,000
Two-seater truck to pull trailer	$40,000
Ten Corriente steers for practice	$5,000
Worthless Blue Heeler dog named Radar	$200
Arena for practicing and socializing	$5,000
Hydraulic chute (cheaper than a divorce)	$3,500
Roping school with National Finals winner	$700
Different roping school with good teacher	$700
Entry fees (to date)	$900

Equipment upgrade:

New saddle	$1,200
EXTREME go and slow bit	$125
Polyethylene urethane no-pressure saddle pad	$125
A box of "no miss" ropes	$250
"Never get'em hurt" horse leg-protection	$125

Image enhancement:

Space-age biothane tie down	$20
Straw hat (came with full-size	
George Strait picture)	$70
Headstall with turquoise	$200

Total estimated expense   **$116,615**

Income:

First in the average at Mineral Wells, Texas	
3:14 p.m., Sunday, May 1, 2006	$228
Picture frame (gift from admirer)	$0

Total income (exact figure)   **$228**

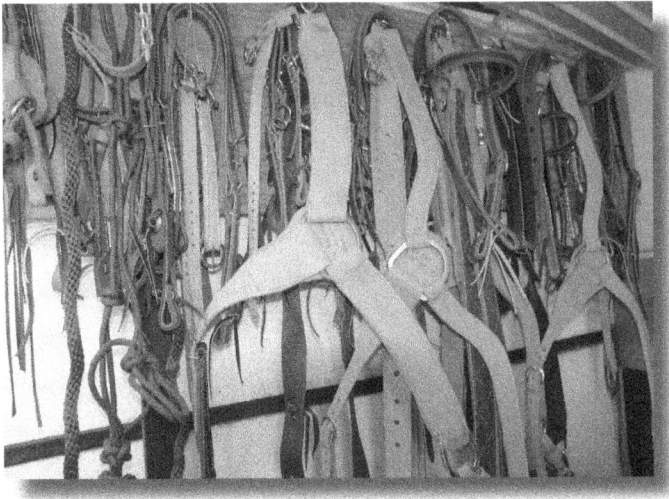

Of course this doesn't take any depreciation into account including the much depreciated wife who tries to keep up with it all.

Roping – what used to be a poor-boy sport no longer is. In Texas, where anything can happen and usually does, the classier covered arenas are now offering golf tournaments in conjunction with their ropings. One arena has a swimming pool, two restaurants, a western store, basketball court and is adding a softball field.

In many areas it is hard to find an arena that you would recognize as such – built with 2" x 12" boards and used bull wire. You cannot go to a roping in an open top trailer, single seat pickup and a ranch horse – you would not be allowed in the gate even if you could whup the entire list of entries. You won't find a single gunny sack girth in the bunch of them.

In spite of the math, every rodeo ground in America continues to be covered over in trucks, trailers, hats, and swinging ropes throughout the spring, summer, fall and well into winter. It's a man's sport, a woman's sport and a family sport. It appeals to doctors, lawyers, a few Indian chiefs and every now and then, even a genuine cowboy.

The used-to-be and has-beens from the bucking chutes find renewed rodeo life at the roping boxes. Where once they chided and scoffed at the "pot-bellied" ropers, they now are them.

Looking for a way to put a little disposable income into circulation? Buy a rope. The rest will just come naturally.

Cowboy Christmas

The long-used term "Cowboy Christmas" has nothing to do with the yuletide season.

This lucrative holiday is for the rodeo cowboys of America. It falls in the summer of the year during the season of the 4th of July. It's "Christmas" because the increased number of rodeos to get to can mean a sizable increase in the money to be won.

Rodeo cowboys commit daring feats of 24 hour-a-day traveling to compete in as many 4th of July rodeos as they can in a 3-4 day period.

It has been said that rodeoing is an addiction and the only cure for it is more rodeo. In two long-running popular songs it is referred to as that "damned old rodeo." Back in the 60s Ian Tyson, a Canadian rodeo cowboy turned singer, penned a song called "Someday Soon."

The song lamented the love a rodeo cowboy has for the sport and the pain it causes those that love him. "He loves his damned old rodeo as much as he loves me." The song has stayed popular over four decades with new recordings of it by Judy Collins, Lynn Anderson, Chrystal Gayle, Suzy Bogguss and Chris LeDoux.

Garth Brooks came out with the next really big rodeo song called simply "Rodeo." The lyrics sum up the sport about as well as any written.

> *Well, it's bulls and blood*
> *It's dust and mud*
> *It's the roar of a Sunday crowd*
> *It's the white in his knuckles*
> *The gold in the buckle*
> *He'll win the next go 'round*
> *It's boots and chaps*

It's cowboy hats
It's spurs and latigo
It's the ropes and the reins
And the joy and the pain
And they call the thing rodeo

Well it ain't no woman flesh and blood
It's that damned old rodeo.

Fourth of July rodeoing brings images of road weary cowboys, tired horses, pickups filled with dirty clothes, fast food wrappers and muddy boots. The dash is full of rumpled programs, Copenhagen cans, empty 7-11 coffee cups, dust covered sunglasses and an assortment of empty CD cases.

Rodeo roots run deep in the heart and soul of the American cowboy. It began as a good-natured competition among the working cowboys. Over more than a century it has evolved to be a major league sport.

Today's rodeo, except perhaps some of the events themselves, resemble little of it's beginnings on the open range. The cowboys have advanced to be true athletes of their events and not always with ranch cowboy roots. The addiction to the adrenalin remains the same as does the dedication to the competition.

And, as I observed most recently, the excuses have maintained continuity through time.

"I started a ride that would have won the whole deal but my rope slipped and then that bull was dancing all over me." We'd have been 6 seconds on that steer if he hadn't ducked his head." "They got me at the gate and I missed marking him out."

The only difference now is those excuses get home sooner than they used to. It is now quite common to see a cowboy standing on the back of the bucking chutes after his "almost great ride" using a cell phone to report to home the details of his wreck.

Now that's progress.

Daisy and Buford...
the Mechanical Mismatch

Cowboys as a general rule are broke so much of the time you wouldn't think they could have many toys. Don't be fooled.

The obvious toy for a cowboy is a rope and the unlearned would be amazed at how many kinds of ropes there are as well as how specialized and expensive they can be.

Then there are the roping cattle. A conscientious cowboy won't rope what is termed the payday cattle (cattle he is raising for market) any more than necessary. Instead he will buy or lease very sorry, contrary minded, expensive Corriente cattle to rope. He'll use them up and then find them unmarketable and not very edible.

The arena can be an expensive little project. It is also mandatory that when there are steers on one end there has to be cold beer on the other. That is an ongoing cost of doing business.

A good cowboy can make a roping horse. That project will require a well-bred prospect and a lot of patience and skill on the part of the cowboy. But sometimes, neither is available at the particular time a cowboy wants or needs a good rope horse. I have lived in houses that are nowhere near as expensive as a finished roping horse.

A roping trailer is a must for the cowboy to travel to roping competitions. The object of the trailer is never a simple get the horse from point A to point B. It seems, and this is his story and he's sticking to it, there are many competition ropings that won't let a man enter if he is pulling an ordinary ranch trailer.

In its stead will be a $35,000 aluminum three-horse slant trailer with living quarters featuring a generator, air conditioning, a full bath, kitchen and a maid. The maid story is for another day. Only a novice roper would attempt to go roping with a trailer that is merely practical.

If the cowboy has experienced a little good fortune and can support his habits to the extent of the high dollar horse and higher dollar trailer, he may consider optional toys.

While never forgetting how broke he is, he will always cut every corner to save a dime. So after acquiring a high dollar electrical roping machine, he fast talks a genius friend into doing the wiring on the new toy called Buford.

Buford is a mechanical steer that operates on a pivot moving at a steady pace in a circle. The cowboy can follow along and rope it, at which point it stops and the cowboy can retrieve his rope.

Now our particular cowboy has Daisy, a good finished rope horse that has won a few saddles and knows her business as well as any. Daisy was the first to try out Buford along with our cowboy. She was cocked and ready to go to work when they turned on the electricity, but something was wrong.

Daisy was in the correct position for her cowboy to rope, but Buford was going around backward. Daisy has seen a lot of steers in her day and wasn't the least bit afraid. She gave Buford the "don't mess with me" eye and pinned her ears to make her point more clear. But Buford kept coming at her.

She is a big horse and refused to give ground. Finally Buford attacked her. Upon impact Buford shut himself down.

Daisy had an ugly cut on her foot, but by golly, she stood her ground. The vet bill was $137.50 and our cowboy decided Daisy didn't need any more training on Buford. We don't know what happened to the genius electrician friend, he's been layin' low.

Gettin' Old Ain't for Sissies

We all get old sometime. It happens to some folks faster than others and cowboys pretty much across the board fit into that category.

The life of a cowboy is hard on their physical body. It doesn't seem to hurt their mind much or if it does, it could be reasoned that is why they stay in the business.

A life of having horses pound you into the ground, cows run you over and assortment of other wrecks involving ropes, gates, pickups and trailers have cowboys feeling some serious aches and pains at an early age. The body is just not made to bend the wrong way as many times as cowboying will make that happen.

By the time cowboys are over 50 they usually look as old as they feel. Bowlegs are just one visible symptom of a much worse problem. Those knobby knees in the middle of that bow are a never-ending source of pain, agony and frustration.

An old but not so-old cowboy with very bad knees wrote this about that situation:

> *Where ever you go either you walk or ride.*
> *You use your knees with every stride.*
> *Your stride gets short and the trail gets long.*
> *It sure is hell when your knees are gone.*
> *You jump right off but when you land,*
> *Sometimes your mouth gets full of sand.*
> *You can't stand up and it hurts to crawl.*
> *You ain't no good on the ground at all.*
> *You can't run your horse with any ease,*

'Cause of the real bad hurtin' in your knees.
But don't you worry about that ol' pard,
The cowboy life was always hard.

With today's technology, more and more cowboys are signing up
for the "spare parts" surgery. Usually these guys need the new
knees long before the doctors think it's advisable because he
knows they will need another set before they are eligible for social
security.

So with that to look forward to, they hobble around dragging a leg,
thinking everything they see looks like it needs to be set on and
giving the anti-inflammatory drug business a dramatic boost in
sales and steady stock value.

At the corral they will look for a place to sit and rest where they
don't have to be tailed back up when it is time to go back to work.
The grunts and moans you hear are just them trying to get their
foot in the stirrup and get back astride their horse. No longer is
there any shame in using a ditch bank, log, rock or trailer fender to
make the job easier.

They find it acceptably easier to the let the "young buttons" do
the work even if it takes longer than it should. With no qualms
they discover a new-found fondness for shorter horses and slower
women.

And those old cowboys that are competitive ropers? They rope
fast because their shoulders won't hold up much past a few quick
swings of the loop.

You will notice most of the "aged" roping events start early in the
morning, and for good reason. Those cowboy diehards want to get
their turn to rope over with before their pills start to wear off.

This gettin' old ain't for sissies.

A New Concept in Team Roping... the HandiCUP Division

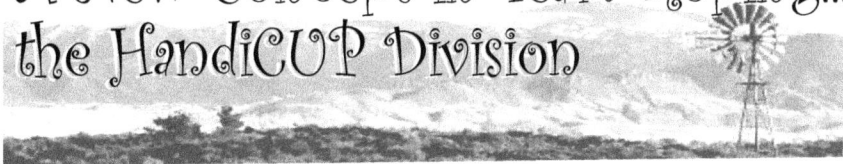

There seems to be an unspoken, in polite company anyway, request for a new division in the world of team roping.

This suggestion came by way of an email dissertation from a female team roper relating the story of her recent return to the roping arena. For those that know me personally, you will understand that I didn't think up this idea on my own because of a complete lack of need for the plan. It would have just never occurred to me.

For you non-ropers, a handicap roping ranks cowboys according to skill level similar to golfers. In an effort to make it fair for everyone, which is in reality impossible, seconds are taken off the roping times before the last go round to even up the field of roping qualifiers. This concept I need and understand.

It is really a business tactic on the part of the roping producer. The more entries there are the more moolah he earns. This handicap system entices lowered numbered (less skilled – that would be me) ropers to shell out more entry fee money because it appears the handicap system gives them a pretty good chance of winning. Fat chance.

The author of this newly proposed concept suggests that in addition to the handicap offered for lack of ability, roping producers would do well to offer advantages to female ropers due to an assortment of extra burdens born by the fairer gender.

She called it the HandiCUP Roping.

Guidelines suggested were that there be a second off your time for being female, another second off for each child given birth to and another second off if you are currently nursing one or more children.

And last but not least another second off for each cup size the roper is endowed with. Proof is not required but the decision of the judge is final.

She swears she never leaves home without her self-designed roping sports bra that keeps everything where it belongs and nothing swinging but the rope.

Now this is the part I'll have to take her word for.

She says, "You have no idea how uncomfortable it is leaving the roping box like a banshee, running nine-oh down the arena, screeching around the corner and then whipping around to face, causing a tetherball-like feeling midsection. The momentum will often take you right out of the saddle and those ropers really hate having to take the time to scrape you up off the ground."

She goes on to say that this handiCUP program takes care of quite a few issues including a get-even to all the men that try to compare the pain of childbirth with cutting themselves shaving. "Not to mention along with the pain the world expects you to 'glow' and then when it's over they hand you a couple Tylenols."

Her belief is that with this system she alone is worth a 16-second handiCUP putting her in the last round of the roping with a negative time and making her unbeatable. "Women from all walks of life will start roping and men will actually want to rope with us. Big bust, many kids – No problem. You'll be the queen of the roping."

I pass on this idea on to the world of roping with the thought perhaps it could be a good idea. What do I know?

It would have never occurred to me to suggest we as females should get any preferential treatment beyond dignified respect when competing in a sport that was first invented for men by men.

But then I've never had much in the way of handicups to get in the way of my work.

Of Duct Tape, Baling Wire and WD-40

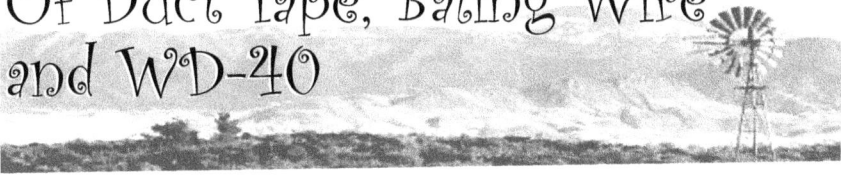

There are some constants for survival that every country girl learns to utilize beyond feminine charm and tough grit.

In the years of hauling down the road – driving hundreds of miles to perform in the rodeo arena for a few seconds and then driving home – I often found a need for basics in emergency management.

Three items consistently required for crisis control were duct tape, baling wire and WD-40. With those items you could fix anything short of an amputated limb.

The duct tape covered an assortment of ills, ranging from the horse's splint boot that wouldn't stay fastened, to the pickup tail light lens that refused to stay in place.

Duct tape could cover a split in a radiator hose, mark the rented stall as "taken" or pad a spot that was poking somewhere or something it shouldn't. I was decades ahead of not-yet-invented Homeland Security in my use of duct tape to seal the ills of the world (road dust) out of the camper.

If the rodeo went poorly, it was also an adequate "For Sale" sign placed prominently across the rear end of the horse that you shouldn't have bought in the first place.

Baling wire is a generic term for any kind of wire of usable size to fix all those things you didn't get around to replacing, welding or repairing in a proper manner. Sometimes called the "poor man's welding rod," no self-respecting horse trailer or pickup truck should be without it for emergency repairs of the wiring kind.

It works well for vehicle mechanical repair (tying up the muffler that just fell off), horse equipment repair (the headstall that broke as your name is being called to compete) and tying the dog to the trailer hitch because she slipped her collar and bit the rodeo clown.

WD-40 literally stands for water displacement, 40th attempt. That is the name straight out of the lab book used by the chemist who developed WD-40 in 1953 when he was trying to concoct a formula to prevent corrosion. The result was a multi-purpose problem solver that has thousands of uses and even comes with a medical warning for those that spray it on their body joints to treat arthritis.

Life is easier with WD-40 around and in those times when the handyman jack wouldn't jack and the trailer hitch jack was immovable, nothing was more valuable than that yellow and blue can of magic.

Unless, of course, you were Barbara.

Barbara, my friend and hauling partner, didn't need emergency backup for anything. She was beautiful, very feminine, with a smile and figure that stopped traffic. Perfectly coiffed at all times, she had long nails, little need for makeup and a natural charm that emanated like sonar.

While I was cussing and beating the lug nuts off a very flat horse trailer tire with all the brute strength I could muster, Barbara would step to the front of the rig and wring her lovely hands in distress. In a nano-second four linebacker-sized cowboys would appear and with a cowboy drawl say, "Can we help you out there little lady?"

Shoving me out of the way like a pesky weed, they would bodily lift the trailer, fix the tire and leave for the rodeo dance with Barbara on their arm like a prize trophy.

I simply watched in amazement and put the jack, WD-40 and duct tape back in the tack compartment.

A thousand times since I have wondered when I missed the "Hand Wringing 101" class.

The Day Buck Won the Big Rodeo

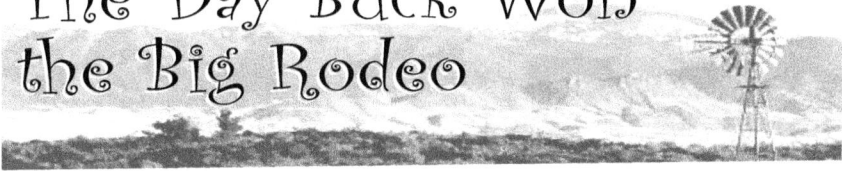

Cowboy stories are shared over and over and usually last through several generations.

As a rule, they either impart a lesson or offer simple entertainment value, or sometimes the stories are an opening overture for a new friendship.

Cowboys value humor almost as much as they value grass, and a gifted storyteller will find himself in demand at about every gathering in the county.

When strangers move into an area, the storyteller is certain to show up to give them his welcome. His real mission is to make them the beneficiary of the wildest of his stories in his repertoire since there is no way for the newcomers to determine any lack of truth.

Such is the story of Buck and Rowdy, which are fictitious names to protect the liars, but the story telling is true.

Rowdy just moved to the county, bought a nice little place, put a few cattle out on grass to make it look right, tuned up his fishing pole, built a new roping arena and proceeded to move into what he liked to think of as semi-retirement.

Buck dropped by one day to help Rowdy out with the Coors Lite inventory in the saddle house icebox.

With the big W's on his Wrangler pockets settled onto an upside down five-gallon bucket, Buck opened the conversation.

"Rowdy, you ever rope any calves?"

Recognizing this as an intro to a story, Rowdy allowed that he had roped a few, way back when.

There followed a Navajo length of silence, just to make sure Rowdy didn't want to tell a story first, then Buck began his story about how he won the buckle at the big rodeo.

"I had been calf roping pretty steady for a good while, but I was always coming in fourth when they were paying three places, or 84th when they paid 83 places," lamented Buck.

He went on to say he had figured out, after giving it considerable thought, that what he needed to win was a calf horse with a real good stop on him. He put out the word and not long after got a call from a fellow he knew. This guy claimed he had just the horse Buck was looking for and assured him that he had a real good stop.

The trade was made over the phone and arrangements were detailed to meet at the big rodeo with the horse. Buck was entered up in the calf roping and when they called his name, he backed his new horse in the roping box.

When everything was just right, he nodded and made a clean break from the barrier. He stood up in his stirrups and threw his best catch'em fast loop.

That was the horse's signal and he planted his backside in dirt like he'd been pole axed. This launched Buck straight between ole "Stop Hard's" ears.

In an effort to save his life, Buck grabbed the rope on the way to the ground and slid down it like a handrail until he got to the calf. He knocked the calf over with his head and while he was in the neighborhood, he tied up three of the calf's legs and threw up his hands.

Turned out that was the fastest time of the day. He won the event, got his buckle and almost enough money to cover the entry fee. He was a happy man.

Out back, behind the chutes, when the rodeo was over and all the other ropers came by to congratulate him and admire the buckle, he managed to swap off that calf horse with the real good stop. That made him a real happy man.

And Rowdy was real happy he didn't need a hard-stoppin' calf horse.

The Smooth-Mouth Team Roping Finals

All it takes for someone to decide there needs to be a new club, association, or way of measuring skill among peers is for a couple of forward-thinking individuals to decide it needs done for an assortment of reasons.

Such was the birth of the Smooth-Mouth Roper Association, at one time known as the old-timers association, followed by a more politically correct name upgrade to "senior" ropers association.

Jim and Ned were genuine highly qualified lifers at ranching, roping and rodeoing. Blessed with a number of similarly-competent and similarly-aged friends, they decided to create a place to compete for only those with their same lifetime of experience.

The consensus was that cowboy attire does not include ball caps, tennis shoes or bling-bling and answering a cell phone while in the arena during competition was not an attribute to anyone's roping skill. They formed this association tailored to the requirements of full-grown men that had lived to see 45 years or older. Maturity was optional.

Familiar with the structure used in an assortment of associations ranging from the Professional Rodeo Cowboys Association to the Sunday school class, Jim and Ned made an executive decision to avoid too many fools making too many rules.

To solve the problems associated with that process they included every rule they had ever heard and then, showing that with maturity comes the wisdom of flexibility, they declared only the rules that seemed appropriate at the time would be enforced.

Jim and Ned gave the new cowboy member, Jesse, the full rundown of rules and regulations including the specified number of approved ropings he must attend. If he placed in enough of them and qualified for the Winner's Ropings he could also qualify for the Smooth-Mouth Finals at the end of the year.

The only other requirement to make a run at the year-end awards was to gather up all the entry fees, roper numbers, fill out the forms and have all the documents and cash to the association office in plenty of time – or he couldn't rope.

Jim and Ned left out the "flexibility" feature of the rules when they indoctrinated him into the smooth-mouthed bunch.

After a successful season, Jesse arrived at the four-day finals to the sounds of the loud speaker announcing that if anyone wanted to enter, they just needed to go to the office. There were extra headers, heelers, extra ropings, and hey, if you wanted to join the association at that time so you could rope, that also would be arranged.

Jesse and his partners gave each run their best shot, taking dead aim at the trophy saddle that was to be awarded to the top ropers. In spite of a necessary rerun because one heeler roped the flagman who happened to be a little close to his work, Jesse was the likely candidate to win the whole deal – and did.

With the flexibility rule in full play, one of the other saddle contenders offered a bribe to one of Jesse's heelers and then was overheard complaining because he was out four dollars and still didn't get the saddle. Nobody said these guys were high rollers.

Roping early so they could compete before their daily pain killers, anti-inflammatories and muscle relaxers wore off, the finals rolled to a close. The cowboys went home with complete respect for one another's competency and competitive abilities, along with the flexibility of the rules.

The only rule Jim and Ned enforced was the five dollar fine for whining. No money was collected for the multiple infractions of that rule, validating just how flexible this association of mature whiners is.

Just Who is the Amateur?

The pickup looked a little "ranchy" and had been re-painted several times. The last time it got a fresh coat, its appearance suggested a whiskbroom had been used to apply the faded blue paint.

The wobbly single horse trailer had never been painted and was complete with wood-slatted sides and the metal bows over the top – no top of any kind, not even a tarp. The gate was crooked and needed wired shut and one could only imagine if the floor was solid enough to hold anything heavier than a small dog.

The oversized palomino roping horse looked better suited to pull a plow, but the poor boy from down on the river managed to use him to compete quite handily in the calf roping.

"Snoopy" he called him. The only explanation he would give was: "Every time someone tells me I need to get a new truck and trailer, I tell them there is 'nuthin' wrong with the one I got.' But, I do worry about my horse a little so I'm in the hunt for some big Snoopy goggles and a Red Baron scarf for him."

Knowing the true path to the pay window, this cowboy didn't waste much on frills but more than paid his way with skills. Pretty is as pretty does.

The other side to that is another story.

It's not uncommon for cowboys going down the rodeo road to pencil their travel plans around a stop at a buddy's place. There, they will run his cattle for practice, eat at his table, and sleep in his bunkhouse – all in the name of a last minute tune-up before the "big one" – rodeo or roping.

It was Wrangler National Finals Rodeo time in Las Vegas and behind every rodeo hand that qualified, there were legions who wished they could.

Some marketing genius decided to offer the "wished I could" ropers a place to compete, close to Las Vegas so the fun and atmosphere was a perk, but the roping was open to anyone with a pulse and check book.

David and his partner, whose name I never did catch, stopped off to tune up for this big "also ran" roping. They arrived in separate $50,000 pickups, pulling $50,000 aluminum trailers and if rig shopping were a contest, they had it won.

The first session went badly. Missed heads, missed heels, missed dallies, missed everything, half heads, bad handle … you get the idea.

After a day, the unnamed header loaded up and headed for the city of lights but David stayed a little longer to perfect his uncountable imperfections.

The first two steers headed for him, David completely missed the heels. No one is quite sure what happened next, but the resident header nodded, roped, turned off and when everyone looked David's way, he was lying in the dirt and so was his horse. The horse got up but David continued to lay there like a dead tuna.

This was not his shoulder's first encounter with the arena floor and David was in bad shape. A bag of ice, some high voltage Motrin and a shot of Tennessee whiskey later, David was absolutely positive he'd never been better.

Assured that his horse was fine, his friends unsaddled him, lifting the gear into the trailer for him because David couldn't lift his arm to poke his own eye.

Asking, "How do I get to Vegas from here?" David waved good bye with his good arm which did not happen to be the arm he needed for roping.

A little detail like a wrecked roping arm that didn't work very well even before the injury wasn't about to keep him from his dreams of winning the world. Or at least seeing Las Vegas when it teemed with cowboy hats, pretty girls and lots of possibilities.

You gotta love'em.

Cowboy Conversation at the Arena

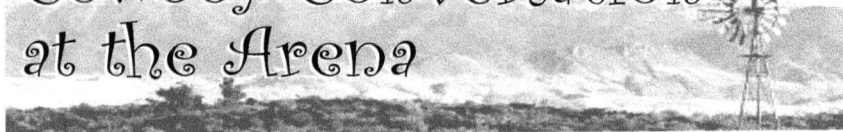

The fine art of conversation is what separates the human species from the sub-human ones – sort of. A cowboy's ability to coherently converse falls somewhere in the middle of the gap.

This took place after an afternoon of roping at one cowboy's arena.

Monte: How about having a barbeque here Tuesday after roping?

Blaine: That would be wonderful. I may be late to rope because my son is getting married.

Dan: Did you see old Slats fly out of that heeling box and get me in position? That is one good horse.

Jerry: Yeah man, Slats really had his game face on today.

Ed (known as Special Ed to his friends): Man, I like barbeque. One time I was in deep East Texas and had the best barbeque you can imagine at a road side stand and the coldest beer I ever drank.

Mark: Would anybody like a beer now? Monte's got some in the icebox here. Who all wants one?

David: Did y'all see my new boots? Ordered these from the catalog, official PRCA sponsored boots. You ought to try some. Make you rope better.

Monte: So, can everybody come Tuesday for a roping and barbeque?

Blaine: That boy ought not to get married. He's too young for that kind of responsibility.

Dan: What's that girl look like you were going to introduce me to anyway? I don't want to get hemmed up with any buckle bunnies.

Jerry: Speaking of buckles, did I tell you all about the time I won first in the average at Mineral Wells and got this great buckle with genuine rubies and all on it?

Ed: Mama used to fix barbeque once in a while. Made the best potato salad. Were you thinking potato salad too?

Mark: This Coors Light is really good and cold. That mountain water sure makes a difference. Anybody need another? Monte's got plenty.

David: That new rope I got really works good. Endorsed by Woodard, and he sure knows ropes. Never missed today all on account of that new rope.

Monte: Somebody toting a purple rope behind me missed bigger than Dallas. Who could that have been?

Blaine: You take your chances toting a purple rope. You know people are on the look out since that movie came out. I guess I ought to be happy my boy wants to marry a woman.

Dan: Slats would buck me off if I even thought about using a purple rope when I was riding him. He is one smart horse and wouldn't stand for that. Tell me about that girl you're going to let meet me.

Jerry: I generally save that buckle for when I'm going dancing. It is one great babe magnet.

Ed: Mama cooked everything good, though. She made great pie. Were you thinking pie too?

Mark: Cold beer don't go with pie. What's the matter with you? If you're thinking barbeque you're going to have to restock this icebox. It's about empty. Anybody want another? We may as well polish off the last of this so he can have plenty of room for more.

David: What ever happened to the old cowboy movies with John Wayne or Lee Marvin? Nobody would have given them a hard time no matter what color rope they were toting.

Monte: I'll be happy to restock the icebox and get the barbeque fixed if y'all will just tell me how many people to fix for.

Blaine: That gal my boy is marrying is sure a looker. If you could get somebody to introduce one like that to you it might be all right even if she was a buckle bunny.

Dan: If I was going to get married I'd want one of those horseback weddings. Don't y'all think I'd look great all dressed up on old Slats? I could try to get my bride a nice horse too.

Jerry: You could borrow my buckle for the wedding.

Ed: What would you have at the reception to eat? Barbeque is pretty messy for a wedding dress.

Mark: I got to go. You're plumb out of beer.

Everybody sauntered out of the barn, loaded their horses and pulled out. David of purple rope and PRCA boot fame was the only slow one and Monte nabbed him before he got away, hoping for a firm commitment.

David shrugged and said, "We'll just have to play it by ear."

Cowgirl Philosophy

It's All in the Jingle of the Spur

Rural dilemmas are usually completely foreign to the majority of the population living urban or suburban lives.

How many soccer moms do you calculate get up before good light, start the dinner makings while she cooks breakfast, turns out the roping steers to pasture, feeds a barn full of stalled horses and then spends a little time throwing a loop at the practice dummy?

All this before the real world is "at work" so she can begin making phone calls to find someone who can put new rowels in her spurs. Losing one in the arena the day before created a near catastrophe.

Finding a spur maker is not a problem if she is native to the area but transplant her somewhere out of her homeland and it becomes a major project.

This is a predicament the world usually doesn't spend much time pondering. But to the world of the ropin' and ridin' folk, a spur rowel is on a list of critical items.

Contrary to what the fancy-dancy cowboy clothing industry would have you believe, it is not a decoration to hang around your neck, but a much needed tool.

Magazine stands are full of slick covered advice on how to expeditiously accomplish housework, cooking, shopping and laundry – all while keeping up on the local and national news in order to be a scintillating conversationalist on all levels.

Even more pages are full of instruction for the wonders of skin care and proper makeup application for the belles of the world to remain beautiful for the men in their life, cowboy or not.

Female health issues garner regular headlines that instruct scheduling mammograms and other assorted tests to ensure healthy longevity and even more pages promise a wrinkle-free face and teeth that glow whiter than those in Hollywood.

These are all concerns that women worldwide face daily in their lives, but the rural girl will find little expert advice on how to quickly locate someone to replace a spur rowel. It is a crisis unique to the lifestyle.

Once located, the spur maker will first assume the broken-spur toting woman to be another Tahoe-driving, Wrangler-wearing wanna-be, the like of which is flooding rural areas of America.

The signature on her spurs will redeem her reputation. Little did even she know that the name stamped in the silver on her spur was one that was legendary to spur makers. She'd known the cowboy to be just one of a fleet of brothers who worked cattle in the Texas panhandle and he was the one that happened to make spurs, bits and buckles.

That moment of name recognition obviously elevated her on the "genuine cowfolk" meter and the spur maker immediately introduced her to his wife, children and all eight of his dogs, each of whom was apprised that this gal owned a pair of Jerry Cates spurs.

This was followed by a litany of rodeo exploits and wild cow chasing stories the spur maker felt obliged to tell to establish his own credibility. The cowgirl just listened as her bona fides had been established by ownership of good spurs.

Cowboy dude designers, including Ralph Lauren, are wasting their time making all that fancy dude stuff to establish the cowboy look. Anybody that wants to be recognized as a genuine cowboy just needs to find themselves a pair of half-worn out quality spurs.

Losing Your First Love

I remember my first love as if it were yesterday. He was tall and handsome. He gently taught me all those firsts with the patience his 20 years of living had given him.

He gave me unconditional love for over two years. He loved me like he loved no one else. I would call his name and he was mine and mine alone.

But then he died suddenly one early summer morning. At the tender age of seven, I learned the reality of losing a love that could not be replaced.

His name was Ranger. He was a dark sorrel gelding that for whatever reason in his golden years, took a liking to a scrawny little girl. I rode him everywhere. I thought he was the greatest horse in the world, never knowing then what good care he took of me. He jumped logs and ditches slowly and carefully enough I thought I was National Velvet.

And the best part was, he'd let me catch him out in the meadow with a small rope and a can of grain. And only I could catch him. I'd call and call and finally he would come.

My dad would try to catch him and he'd run 'til my dad was mad enough shoot him. If Ranger needed caught for anything, I had to do it. I'm sure it was the very foundation of any self confidence I was to gain in life. He made me feel very special.

As I grew up, I had many other horses, but none as special as the first one, Ranger. He found a place in my heart and soul that will never leave me.

Last week those emotions were stirred again when I watched my son grieve the loss of his first horse. Hoot, another fine old warrior who had raised a couple other little boys and then came to raise mine. He apparently liked little boys as he did such a good job letting them think they were in charge.

Little cowboys are pretty big in their minds at a very young age. You'll see a three year old pull his hat down tight, buckle up his chaps and insist that he can rope anything that needs roped. If Dad can do it, so can he, just ask him.

Hoot had been a calf roping horse in his younger years. When an adult rode him and took down a rope they better have a deep seat because Hoot was going to go to work and do it at full speed. Crossing rough country with ditches and deadfall, Hoot took care of himself and the adult rider better just settle in for the trip.

But, not with the little cowboys. He almost tiptoed through the brush, gently stepped over the wash outs and if a rope showed up in the hand of a kid, it was like he didn't even notice. He nurtured a little boy's confidence with every ride. And he did it until his age no longer let him safely travel the pasture miles.

When old age finally took the old guy it was a blessing for him, but a sad day for the cowboys, big and little. Hoot, you won't be forgotten. Little cowboys growing up to be big cowboys will always remember their days in the saddle with you. Adios old friend.

The New Face of Ranching

The reality of genuine cowboying in today's world isn't a pretty picture and the future of it is changing fast. Ranches that actually need hired cowboys are dropping in numbers faster than the temperatures in January.

Pressures of government regulation, environmental restrictions, high costs of doing business, and high dollar opportunity from land developers are just a few of the factors working against the longevity of the ranching industry.

Still, almost every little boy at one time or another had cowboy dreams. Memories of their favorite stick horse, a set of plastic six shooters and a hat trimmed in red with a stampede string on it are planted firmly in the imaginations that pinned on a badge, chased the bad guys 'til sunset and roped wild cows 'til suppertime.

And those little boys? Many of them grew up to be big boys who still want to be cowboys. They've gotten college degrees, work in big business, hold down suit-and-tie jobs and manage a portfolio that may or may not let them afford to be cowboys in one form or another.

It is those "cowboys" that are buying up the West and fantasizing over their new-found careers. While they are warm and safe in their southern climates waiting to sign the papers on their dream in the West, they have no idea it is taking snowmobiles to reach their new ranch headquarters. Their mental image is the same as they day they looked at the place on a summer day – green pastures, summer Aspens and trickling brooks full of trout.

Their first move will be to buy some cows to stock their ranchette.

Tradition will have them overstocking with cows that forgot what it was to chew with teeth or young heifers new to the business of birthing babies.

And of course to have calves, you need bulls. Even a pilgrim knows that. So they will buy the biggest one they can find because he's "so pretty" and partner him up with a fence-jumping longhorn bull to do the job. They have no idea the hell they have just put into the business of cattle reproduction – for them and their neighbors.

When the grass is "sheeped-off" and the cows go back to the sale barn, the ranchette will be fully stocked with one paint mule and an emu. In addition, of course, is the ranch remuda consisting of an ill-tempered Tennessee Walker, a very old, very fat mare and one idiot Appaloosa, all with plaited tails and fancy horse blankets.

A wide assortment of hats will be seen along with boots of every nature from mukluks to combat types. The new "guys" will have a cowboy vocabulary no one ever heard but can be traced to old Randolph Scott films.

For many, ranching through the generations is coming to an end that sounds much like a slamming door on an empty tin building. The echo resonates over the land with a hollow ring to it and fades on the horizon.

The ache in the heart and the grip in the gut make no noise as the last generation pulls down his hat and walks away from the land that wore his name for a hundred years. That, sadly, is the new face of ranching.

Salute to the Feed Pickup

As the calendar changes from May to June and temperatures rise from hot to hotter, summer looms before us. Before I bid goodbye to winter, I want to give recognition to a tried and true friend, the feed pickup.

"The Feed Pickup," in whatever shape it takes for each outfit, has long been the most underrated "cowboy" since supplemental feeding began! This beast of burden can single-handedly move more cattle with less trouble than a bunkhouse of buckaroos and band of barking dogs.

Now I'm not trying to slight the ability of cowboy or canine, but sometimes less is more. At the primal urging of food, cattle can be led like the Pied Piper led the rats to the river.

Mom and Pop operations find the art of trolling with a feed sack invaluable. How else could one alone move a whole herd, nine or 209, to a trap for tomorrow's branding?

When the hired man doesn't show up and the neighbors are gone to town, what other way can Grandpa and Grandma get the heifers to the calving pens before the storm hits?

There is a new drinking tub on the far west side of the six section pasture. A sure way to bait the bovine to a new water source is with the sound of the feed pickup horn and the rattle of the handle on the cube feeder.

Feed pickups come in all shapes and sizes. Geography dictates the types of feed. In the North Country where hay is fed in bales or loose, appropriate feeding vehicles are designed to roll on wheels or slide on skids on the snow.

Motor driven or horse drawn is also a matter of preference and ability. In my neck of the woods not much hay is fed, but winter brings routine block or cube feeding until it rains and often that is not until July or later.

Large overhead bins store the cubes brought to the ranch a semi-truck load at a time. Cube feeders are mounted on the back of any stout, preferably 4-wheel drive vehicle the ranch can spare and that the wife can manage to drive. It's been proven that power steering is not mandatory but preferable if the Little Woman is to maintain a pleasant state of mind.

Sometimes folks "honk'em in" with regular pickup horns. When those wear out, the rancher will resort to his own vocal chords. The sound will vary from man to man but the bellow released would give Tarzan some stiff competition.

Some will call the cows to them with a siren that can be heard in echoes through the hills. We hook an air horn to the propane that fuels the pickup and in the middle of nowhere it sounds like a convoy of semi-trucks in a traffic jam.

But the cows know the sound and they will come on the run. Even the neighbor's cows will line the fence with a hungry look. The sound of food in cow country is universal.

In a world where being a cowboy has been emulated and romanticized far beyond the actual job that it is, somehow the endless laurels of the feed pickup are never mentioned. It doesn't eat unless it works. Unlike the wife, it doesn't whine when it's cold, well at least only until the engine warms up. It becomes familiar and comforting to a new mama cow in the spring.

She'll follow that pickup without question but let a man on horseback show up and she's headed for tall timber and not coming back.

So here's to you Feed Pickup, the unheralded hero of the plains. Though your likeness will never grace the walls of galleries next to the works of Remington and Russell, to the everyday cowman you are the backbone of his winter operation.

Storms Never Last Do They Baby

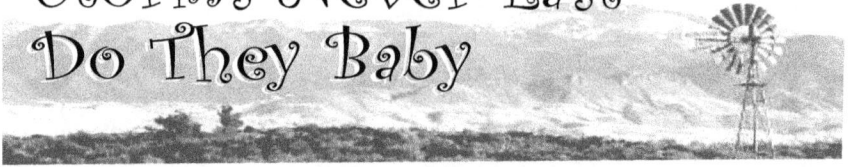

If the title lyric is tickling your memory banks, it is because it isn't original. It belongs to Waylon Jennings and Jessie Coulter. But the memories are mine.

There are many ways to take nostalgic trips through the recesses of your mind. Some work faster and better than others, but it's a journey we all need to take from time to time.

Flipping through old photo albums is a surefire way to bring back flashes of the past when people were younger, thinner, and married to somebody else.

There is something better for the journey than being faced with pictorial proof of how different things are now. That vehicle is music.

Music is a venue of travel into yesteryear that seems to evoke more emotion and less reality than old photos.

An old familiar melody has a way of quickly reaching inside our hearts and souls and touching feelings that, long ago, were pushed behind us as we dove headlong into life.

Lyrics might tell a story that was your story and in doing so, they speak for you.

A song might bring recall of parents dancing to the melody as it played on the radio or at an old country dance that gathered people from the hills and vales to socialize in hard times.

I can promise you, I'll never, ever hear Hank Locklin sing "Send Me the Pillow That You Dream On" without remembering my grandmother's caterwauling of those lyrics as she went about her daily chores.

Musical memories are never better than when they are shared remembrances. They don't have to be specific but memories of an era can bring people to common ground. People that didn't share directly in your life 30-40 years ago, very likely heard the same music you did.

Circumstances may vary but the emotions that erupt do not.

Nothing will turn back the hands of time like an evening with some heart-tugging, boot-scootin', make-me-think-I-can-sing music to put you over the top.

When the fiddle bow strikes, the guitar strings hum and melodic voices fall into a lyrical tour of fine old country music, today ceases to exist.

For a space of time, the room becomes a world of its own in a place a long time ago.

Whether you are swaying to Lefty Frizzell's "Waltz of the Angels" or hearing the alluring Bob Wills' "Faded Love," something begins to happen to your heart.

Traveling down the melody lane, you arrive at a place in the recesses of your mind that everyone should visit. Whether it is the lyrics, the melody or both, something begins to chip away at the shellacked veneer that day-to-day living paints over us for survival.

The musical map carries you forward through decades of "classic country" that becomes a slide show of loving, laughing, crying, and dying.

Waylon's words are immortal in many lives and certainly in mine.

Storms never last do they baby. Bad times all pass with the wind. Your hand in mine stills the thunder. And you make the sun want to shine.

I'm very glad I got the chance for him to remind me of that.

The First Liar Doesn't Stand a Chance

All cowboys are familiar with the "first liar" precept. It is a solid truth that the first liar doesn't stand a chance.

While holding a horse for the horseshoer, a cowgirl was conversationally filling in a gap in the ongoing, visiting with a story about her bay horse who was a wheat pasture mount by trade.

Her husband had used the horse one day to doctor some yearlings. Somewhere in the pursuit of roping a sick one, the bay decided to buck the husband off. By the time the cowboy got up and dusted his britches off, the bay had run the steer down and was standing next to him, waiting.

Then it was Jimmy the horseshoer's turn. With a look of pity at the cowgirl, Jimmy told his story.

As a kid, he had owned a horse that could, he claimed, sort cattle by brand with no help from a rider. His family had been on a ranch that joined a pasture where bucking bulls were kept with the regular ranch bulls. The owner would regularly put all the bulls up in a corner, sort off the rodeo stock and take the bucking bulls to the rodeo.

Jimmy said when he turned his horse in with those bulls, he would sort them all by himself. That was the end of the lesson. The cowgirl never again told the first story.

Cowboys have a knack for taking any story and making it their own. If this one sounds familiar it likely is, but it was told to me with all the sincerity of a true story.

Jerry, Wayne and Tommy were all headed to the big roping on a Saturday morning. Since Wayne's truck was the one that was running that particular week, they all loaded up with him.

Jerry and Wayne are big cowboys by most standards. Both were broad shouldered and stout enough to work hard and not take any guff from anyone. But Tommy, on the same scale, was a giant compared to his buddies.

Tommy was also the wisest. When they loaded up in the pickup, Tommy got in the middle just in case there was a gate to be opened on I-40. It's the old "the real cowboy sits in the middle to avoid opening gates" trick.

So with Wayne driving, Tommy in the middle and Jerry riding shotgun, they headed down the road. They had made room in the floorboard for the cooler and Jerry was drinking a screw-off lid Coors and tossing the cap around to entertain himself.

With the truck seating arrangement causing an uncomfortable coziness, Tommy in the middle would put his arm on the back of the seat behind Wayne to make a little extra room.

Miles passed as they gave each other pointers about roping, women, and life in general.

After a while Jerry began intermittently dropping the beer bottle lid and then bend down to pick it up off the floor of the pickup.

Wayne noticed the big rig truckers were honking and pointing at him as they passed and he finally put it together.

Every time a big rig would come up on them on the interstate, Jerry would duck down after the beer cap, leaving Wayne and Tommy all cuddled up.

All those truckers could see was Wayne behind the wheel with Tommy next to him with his arm on the back of the seat behind him.

He calmly told Jerry that if he dropped that cap one more time he would stop the truck and a genuine whuppin' would commence.

Cowboy stories are as much part of the culture as the horses they ride and spurs they wear. But the lesson to be learned is simple. The first liar doesn't stand a chance.

The Little Woman is a Mighty Force

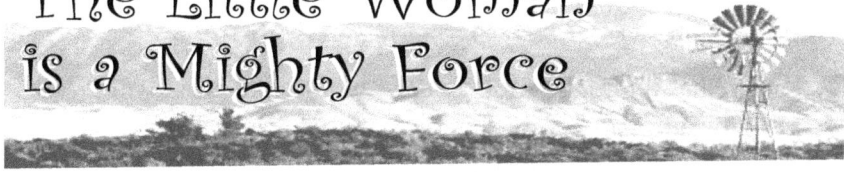

I feel certain that the saying "behind every good man is a good woman," was born in agriculture. Quite possibly it dates back to when agriculture was in its earliest crude form of poking a stick in the ground and dropping in a seed of corn.

Today agricultural methods are not the same but the role the Little Woman plays hasn't changed much. She picks up the slack wherever there is some slack to be picked up. Another phrase born of her schedule is "a woman's work is never done."

Women in general are a tough lot but for the purposes of this pat on the back to the Little Purty as she is often referred to, I'm narrowing the topic to ranch women.

A ranch wife will leave at daylight with the boss, after breakfast is cooked and the kitchen tidied up, horses saddled and lunch packed. Many of those days there is no lunch because the all day event was one of those "it'll only take a little while" projects. They will be horse back all day fighting cows and elements and all the things that can and do go wrong, will.

She will find a way to shut a barbwire gate that only a he-man body builder would be able to pull up to the gate post. She will sit and wait patiently for hours on end right where he told her to wait. Then she will find out he expected her to read his mind when he changed his mind. She'll take a cussin' when it should really have been for the cow that ticked him off in the first place.

And when it's all done she'll come home to a cold dark house, build a fire, fix a meal and get ready to do it all over again. Somewhere in there gets squeezed the grocery shopping, the laundry, house cleaning and bill paying. That has to be done in her spare time, whenever that is.

When she has an important event she would like him to attend, he's always got more work that he can handle and just doesn't have time. So she rises to the task and helps him get the work done so they can go together to something he didn't want to go to in the first place. When he has an event and she's busy, he goes alone.

She may only be five foot tall but stature has nothing to do with guts and grit. She'll sit in a saddle in a blizzard helping to drive a cow to gate that not even the cow can see. After it's all done she'll coil up her wet freezing rope, peel the ice from her batting eyelashes and tell him how much she loves working with him.

She would love a thank you and pat on the back for a job well done. But most often he just thinks of it as getting done what had to be done.

He won't stop and ponder the fact that the Little Woman is the best help he has because she is, more often than not, the only help he has.

The Siren's Song of the West

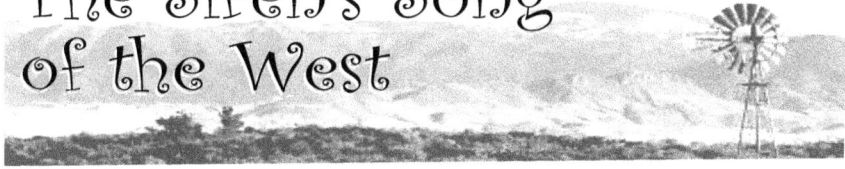

It is a song not audible and yet it pierces the heart of men in every walk of life.

Like the music of the mythological being, the siren's song of the West pulls, tugs and creates within men an unexplainable desire.

It calls them to a way of life in a place where renewed hope springs eternal and they believe for a better life in a less cluttered world.

The sirens of Greek mythology lived on a rocky island in the middle of the sea and sang melodies so beautiful that sailors passing by could not resist getting closer to them.

Following the sound of the music, the sailors would steer their boats towards them or jump in the water to get closer – both ending in disaster on the rocks.

Horace Greeley, has been credited for popularizing, 150 years ago, the idea of "Go West, young man, and grow up with the country." Today, the West is still a magnet to men and women of all ages.

A study of Western culture revealed three out of five men and nearly half of women would like to be cowboys for at least a day. Many have opted for complete lifestyle changes.

In droves, they have packed up their lives and moved to the West, finding a place in the open spaces much like the 100 years of homesteaders.

The 2000 census showed eight of the ten fastest growing states are in the West, led by Nevada.

Not too long ago, 1,200 Michigan residents stood in long lines eager to head for Wyoming's rugged, cold terrain answering a call to a job fair.

The sheer numbers dictate that not everybody can be a cowboy. But a good number will take on the trappings of the trade, buy a 40-acre ranchette, and put a rocking chair on the wrap-around porch to watch the sun set over a small barn that houses two horses, a 4-wheeler and a couple of llamas.

It is a new West and is clearly an amalgamation of the many phases of an evolving genre.

While the West does not own the cowboy, it is the cowboy that epitomizes the West in the minds of those that seek him.

Some men are born to ride and some men were born to sit in traffic. Some come to live in the West as it is now, with a more modern version of the cowboy wearing sponsorship tags on his shirt and making a few hundred thousand dollars a year riding bulls or roping calves in the rodeos.

It is a West where cattle are still king and four door pickups and aluminum trailers ferry the cowboy crew miles across ranches, counties and states – a West where ranchers hang on to an ever-changing way of life, necessitating better practices in order to stay on the land.

There are those who come to feed their soul from the history created by those who came west to grow with a new country.

These were men who rode hard, shot straight and died young. Their ghosts walk the boardwalks of old towns in western territories and call to a breed of modern men who find themselves living a century past their time.

While the siren of the West may not lure man to disaster, the man that heeds the call will find today's cowboy life is not in the clothes he wears or the substance of his dreams.

To this day I have not ever seen the visiting pilgrim come to the ranch, dressed out in his version of cowboy clothes, begging the boss to let him drive the feed pickup.

Now there is a sign of a complete lack of understanding about how the West is really won in this new millennium.

I'm a Ranch Pickup

I took the online test to find my "true inner vehicle." It was one of those goofy 15 question quizzes that lead to a personality synopsis telling me what kind of "vehicle" I would be if I were a car.

The option wasn't there but should have been. I'm sure I am a ranch pickup.

You know the kind. Not much to look at but it will get you there. It might choke and gasp a little, but it will get you there. It pulls to the left and has a front end alignment problem but it will get you there with a tired right arm.

The fenders don't match, the windows don't work. It's not classy, not elegant, may take a little "herding and handling" to go the right direction, but it'll get you there.

Probably needs some engine work and definitely needs new brakes. The A/C isn't all that good anymore, lots of hot air.

Tough, dependable and functional. The kind that can go through a wreck and look the same as before the wreck. Yes, my inner vehicle is definitely a ranch pickup.

There was a time when driving a pickup wasn't the status symbol it is today. Every ranch woman longed for a car to drive to town. If you were rich, you might even have a Cadillac.

Today the same money will buy you a Cadillac, a single-wide mobile home or a four-door pickup.

Pickups are now called trucks. Back then, a truck was those big things that had the word "semi" in front of it. but we've evolved. They have advertisements that proudly tout the roomy space for five passengers. I remember the days when three adults and four kids rode in a pickup, all in one seat and the guy in the middle did the shifting.

If the radio worked, the driver was in charge of the dial. It was also an opportunity for a conversation. We would actually talk while driving the road. Now the back seat of a pickup affords a view of a television screen with a DVD player and the front seat has a AM/FM radio, tape deck and CD player. No one talks.

Those ads over the past 20 years have been quite effective. People have never seen a dirt road drive, mega-diesel engine, four-wheel drive truck, or have them custom detailed, and listen to computers tell them when to fasten their seat belt, change the oil, and fill it with fuel.

The spare tire is now "handily" under the pickup where nobody but a scientist and a linebacker can figure out how to get it off. And walking five miles for help is easier than getting the sissy little complementary jack out from under the seat that is, of course, loaded with groceries, kids, parts and a week of accumulated mail.

Then, most certainly in the dark on the side of the road with a flashlight in your teeth, you put that handily engineered rod through a little hole handily located next to the license plate hoping to connect with the handily located crank on the apparatus that handily holds the tire under the truck.

Changing a tire on today's "trucks" has caused more people to lose their religion than anything you would usually associate with sin.

It is a great trip down memory lane to the days when we actually drove 55 mph, had no A/C so the windows were always down and no lights so we had to be home by dark.

The stick shift and lack of power steering precluded the ability to talk on cells phones, put on makeup and check email on a laptop, all while driving.

My inner self that is a ranch pickup is most definitely one of those original models. Almost a collector's item I'm sure.

Barkalounges and Pawdicures...
a World Gone to the Dogs

It was in 1982 that Hank the Cowdog first showed up on America's bookshelves and now, six million copies later, there are very few folks who haven't at least heard about Hank.

For those of you that perhaps lived in a cave somewhere and don't know Hank, he is the creation of Texas-born cowboy turned author, John R. Erickson.

Based on actual dogs Erickson once worked with at the ranch, Hank the Cowdog is a scruffy, smart-alecky super-sleuth with a nose for danger and an eye for the ladies. And as "Head of Ranch Security" on a West Texas ranch, he's usually up to his ears in all kinds of amusing trouble.

Whether he is called upon to bark-up the sun, investigate suspicious goings-on, or defend the ranch against marauders, Hank's hilarious, hair-raising adventures delight young and old.

Hank, his timid sidekick Drover and his sworn enemy Pete the Barncat have shared adventure after adventure at the M Cross ranch through soon-to-be 48 books.

To know Hank is to love Hank. He is the epitome of the typical ranch dog. He represents the thousands of his brethren lying in the shade of the gas tanks, under the porches and behind the stock trailers of America's ranches.

Hank and I have a lot in common beside the scruffy, smart-alecky, and laying in the shade parts. We are just who we are. The ranch is where we want to be and assigning ourselves jobs that surely make the world a better place is what we do. I don't bark at the sun, but I have barked a few varmints into a corner, usually two-legged ones.

I have often wondered what it would be like for the Hanks of the West if they were to visit one of those fancy city salons for pampered pooches. I've been told about a couple with intriguing names. One is called The Grand Paw and the other The Frou Frou Room.

For the pampered canine, these "kennels" offer overnight suites and plush cabanas, a training and agility course, oral hygiene, organic and holistic foods, a grooming salon and a day spa that include milk baths as well as hot oil and silk treatments.

For DOGS they offer aromatherapy, massage, birthday cakes, pawdicures, a day camp or social club, and an indoor barkalounge (as opposed to an outdoor one, I suppose). Of course exercise, affection, a chauffeur service and doggy psychology are also available.

This is a world where the dogs have names like Ambrose, Guinvere, Tranquilla and Nicolette instead of Gus, Slick, Murphy and Damn U.

Hank would be amazed at the money spent on dogs. Fifty dollars a night can buy you a pawdicure and time in the barkalounge (whatever that is!). Hank and I both know fifty dollars a day in our world is called "wages" and a barkalounge sounds very much like a cowboy honky tonk.

And about Nicolette. Hank would like her to know that if she wants to visit the ranch, he'd be happy to give her a tour of the machine shed and the post pile. If she was real adventurous, he would even take her out to meet his coyote buddies, ole Rip and Snort.

Nothing says cowdog romance like a night howling at the moon.

And Then There was the Pickup Truck

Nothing will spawn a parking lot full of pickup trucks faster than a free meal or the promise of a good "bull" session. The hope for both is even better bait.

It is always interesting to me how you can "read" the group gathered within by the vehicles parked outside.

Recently I attended a luncheon meeting that assembled at a swanky county club by a lake and golf course. The four middle rows closest to the door were full of bumper to bumper, door to door gray, white and the occasional Forest Service green trucks. "Yep," I thought, "free meal."

The occasion was a gathering of an assortment of government agencies all tied together by a weaving of entities only the government could create and no one understands. And they all drive pickup trucks.

As I parked my truck, I glanced across the parking lot to discern if there was possibly some golf bags discreetly placed in any of those vehicles for later afternoon recreation. After all, it was certainly a convenient opportunity.

While I didn't see any golf equipment, I did notice that most of those vehicles were fairly clean and certainly cleaner than mine. I know some of them have to, on occasion, drive down a dirt road to government projects. But it appears they find a car wash more often than some of us.

Every year in December, a gathering of hundreds of ranchers and other ag related people assemble in a big fancy hotel in

Albuquerque for their annual "stockman's convention." Now there is a picture of glaring contrast between urban and rural.

Although the hotel still draws its usual crowd of fancy car drivers, they are only a peppering of class throughout the overflowing parking area. Towering above the Cadillacs, Lexus and Lincolns are row after row of big heavy duty four-wheel-drive three quarter or one ton pickups in both the standard and flat bed versions.

If an onlooker has any sense of the rural world in New Mexico, they know these are bona fide ranch trucks. A closer look will reveal a selection of mud in colors to match the geography of the originating ranch. Ropes, chains, handyman jacks, tool boxes and even an occasional errant mineral block yet to be fed to the cattle adorn most of them.

Big dirty pickup trucks in large numbers. How can I convey to the average person what a sense of comfort that sight brings? It is like arriving home in the dark to see the lights in the house turned on, feel a warm fire and smell a cooking pot roast and baking bread when you step in the door. That kind of assurance all bundled up in the metal of a dirty pickup.

This particular parking lot, or any one looking just like it, tells me that gathered inside are large numbers of the very heart and soul of the land we walk on. Within those walls are fourth and fifth generations of families who endured hardships that we can only imagine to make the land their home.

And for as long as those ranch pickup drivers can still assemble to plan the future of agriculture, there will always be hope for the future.

They can't make it rain and they can't control the markets. But they can and do encourage and share their grit to endure. It is a birthright for each of them.

Dirt Roads, Rough Hands and Sweat-Soaked Stetsons

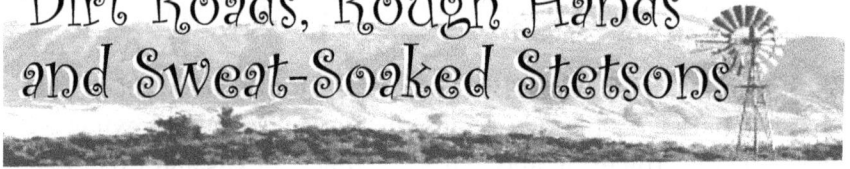

Something wholesome radiates from a person who lives in the realm of cowboy.

Show a photo of a young rodeo hand or a seasoned veteran of the cow wars to a city-folk type and one of their first comments will be to point that out.

There are a number of things that promote the wholesome image of ranch and rural living. Some of them come with their own deeper meanings of life and have infinite depth beyond face value.

A few of those things are dirt roads, rough hands and sweat-soaked Stetsons.

Dirt roads lead to good things. They slow down life and often end at the open door of a welcoming neighbor.

They signify a way of life that has not yet fallen to the asphalt and concrete of a white-collar world.

I have lived down a dirt road most of my life. It is a world unto itself no matter what decade it is. Weeks are without weekends as everyday is the same.

Childhood memories are of endless summers with homemade ice cream, digging for fishing worms and camping along the creek.

When I turn down a dirt road headed to anywhere, I get a "right at home" kind of feeling knowing when I get to where I'm going, it'll be good. A dirt road drive is often a step back in time.

Rough hands of the men and women who work on the land command a deep respect. Those people come with a firm handshake and wisdom born in the sweat equity of life.

The calluses are badges of determination that tell a story matched by the lines around the eyes.

Years of physical work and suppressed worry leave their mark.

There are truly fine people attached to those hands that could sand a board smooth without sandpaper. Like their hands, they are hard as steel at first glance but found to have a gentle nature within. The burdens of life have been worked out through their hands.

And those sweat-soaked Stetsons – that band of dark dirty grime that builds up at the bottom of the crown and spreads out onto the brim – is a cowboy's emblem of never-ending toil.

My dad wore what was my first memory of that icon of the West. Some years after he passed away I wrote a poem about him and included mention of his hat that was so much part of who he was.

He lived in the days when a contract was a man's handshake.

Too far to town, so you made do with what you could make.

Denim shirts, bags of Bull Durham, and rollin' your own.

A sweat-soaked Stetson, shotgun chaps, and a saddle were home.

Those toil-marked hats come in many shapes, colors and sizes. When they have reached the sweat-soaked stage, they take on a common out-of-shape look. They have creases and curls where there should be none and they droop in places not intended for "style."

Often they have a hole or two rotted completely through the brim or the crown.

They wear a little windmill grease, manure and a few blood spatters from a long ago cow-in-the-chute incident.

As time goes by, the hat uncannily takes on an appearance that very much matches the personality of its owner.

Dirt roads, rough hands and sweat-soaked Stetsons – all things so very much more than just what they are.

Old Fashioned Therapy at the Woodpile

As the temperatures drop all over the country, except for maybe the Padre Islands or other like-tropical spots across the banana belts, homes are being prepared for winter.

There is still a lot of the world I live in that burns a wood fire to keep the frost off the furniture. We have an abundance of trees, legally designated for clearance, that aren't good for much else except depleting an already depleted water supply. So it is a double good feeling of warmth when that cedar fire puts off a glow of BTUs.

Without going in to the history of fire and the first guy with a wooly mammoth breech cloth striking two stones together so his woman could charbroil a haunch of deer, it is a fact that fire has been at the center of our culture since it was discovered.

At the same time, fuel and its sources are equally as critical to sustaining fire. The settlers that settled on the plains found their fuel would be supplied in the form of dried grasses, roots and buffalo chips.

But those that sought the realm of the forest found themselves with the blessing of trees. That wood fuel, over generations, would become a source of mental health, physical well-being and an occasional child-rearing tool.

A hundred years ago, 40 years ago, and sometimes even today, the woodpile is not just the woodpile. As often as it has been the place firewood is stored, it also has been the center of an assortment of therapies.

Every rural child I ever knew dreaded the "trip to the woodpile." For those of you that missed that life-defining tactic used by your dad to talk some sense into you, let me explain.

The woodpile was just far enough away from the house your siblings could see you getting the riot-act read by Dad, but far enough away they couldn't really hear. But they did watch to see when the butt paddling came that ended all verbal conversation.

When tempers flared and life became heavy with emotional worry, cords of firewood that needed split were the sure cure to what ailed you. With a splitting axe swinging overhead and the loud crack of the wood as it fell to both sides of the splitting stump, the rhythm of the sound told you how far along the therapy had progressed.

Let a teenager back talk his elders and he was soon building Mr. Universe muscles swinging that same splitting axe. The treatment was likely to last all winter with the pile of un-split wood mysteriously growing each day as dad wore a determined grin on his face.

It doesn't have to be winter to use the "out behind the woodpile" method of repentance. There is nothing, absolutely nothing, that will better get the undivided attention of a youngster who thinks he knows it all, than some dedicated woodpile time.

America could use some more woodpiles. We have lost much of the core of our moral fiber with the flip of a thermostat control. When our government was made up of men who had done a reasonable amount of time at the woodpile, things were different.

In this era of political nonsense running amuck, I propose there come a time very soon we look for that man who knows how to use a splitting axe. He is bound to have built some foundational character out back of the woodpile.

Down Dusty Roads

Short Pay and Fast Horses

There is an old phrase used by the cowboy set, *"ropin' for short pay,"* which basically means long hours of hard work for not a lot of money.

Since roping was, and is, just a small part of the job to be done, short-pay wages covered all the work that needed done.

It was in the beginning, is now, and ever shall be the case for doing the work of a cowboy.

I broke into the world of a paying cowboy job when I was 15. My brother and I hired on for a summer of riding pastures checking yearlings on a high mountain Colorado ranch, which also happened to be our home.

Since I was the oldest, I got preferential pay of fifty cents more a day than my 13 year old brother. I was reaping in a big $5.50 a day while he had to settle for $5. In my mind, that made me the "girl" in charge. He fell for that most of the time.

About 4,000 yearlings arrived in May and came from a much lower altitude. The ranch pastures ranged in altitude from 7,500-9,500 feet or higher in some places, and those critters managed to climb to the highest.

At that altitude, cattle often develop what is called Brisket Disease, Mountain Disease or High Altitude Disease. The animal will develop edema in the brisket along the neck to the jaw or the underline of the belly before it dies. Early detection is the only hope of saving them.

A daily check and count of every single animal was necessary and an accounting of the dead was a must for the record books. That was our job – to look at and count each one in all of the pastures we were assigned, bring in the sick, and cut the brand off anything we found dead.

We would leave early, often with a sandwich rolled up in our jacket on the back of our saddles, and hope to be in sometime in the afternoon before the late day rain showers.

We took turns opening gates as long as were we getting along. It wasn't uncommon to say "we'll meet back at this gate in an hour" and whoever got there first would go through the gate and wait on the other side, refusing to dismount and open the gate again for the late arriver. Excellent fodder for a teen shouting match.

Looking back, I'm thinking our short pay was probably due to two things. Economics of the times was one. I think Dad was managing the ranch for about $550 a month and raising four kids on that.

The other reason I surmise was our youthful unreliability. We got the job done eventually. However, the days were interspersed with opportunities to go for a swim in a pond if it was hot. Often there was a horse race when we were sure no one could see us racing – a forbidden sport.

And about every other day there would be a knock-down-drag-out fight over his roping everything that didn't move and then me refusing to give him the head count since he was busy playing.

Fortunately, I rode very fast horses. It saved my life on more than one occasion.

As the Antenna Turned

There are a couple generations still around who remember when life was simple.

You know – the litany of basics kids today hate to hear us talk about because they have no clue what we are talking about.

One of those things is the TV antenna. You will remember that it never stayed pointed in the right direction to get a clear picture – if ever you got a clear picture. I grew up thinking it "snowed" on the Ponderosa every Sunday night.

A recent discussion about the things individuals once braved in regards to the television antenna invoked memories of those simpler times.

Urban dwellers learned the fine art of rabbit ear adjustment including additional enhancements such as tin foil strategically placed.

Other adjustments could have included an additional wire run from the "ear" to a window screen or metal window frame. Was tin foil really tin or did we just call it that? And when did it become aluminum?

In rural areas, reception required an exterior antenna and usually the higher off the ground it was, the better the reception.

It often was several hundred miles to the nearest point of origin for the signal that brought one channel and a few favorite shows in black and white.

The common denominator for those antennas of old was the pipe it was mounted to in order to reach such heights and the fact it required regular adjustment by turning to bring it into signal alignment.

As a kid at home, one of four, the drill for us was to climb a ladder, shinny up a roof corner gutter and turn the antenna mounted on the peak of sharply angled tin roof. Someone had to stand at the back door and relay the status of the effort. "That's good!" or "Turn it a little more. No, no, go back a little. You passed it." And I might add we all lived to tell about it and not one of us ever fell off the roof.

In the aforementioned discussion, the use of channel-lock pliers or a pipe wrench for implementing the turn was a common practice. Many made claim to a luxury version of the rooftop set up with the antenna placed in a pipe set in the ground and accessed through a nearby window, avoiding the rooftop climbing adventure.

Inclement weather was a given when it was time to turn the antenna. Wind was the usual culprit to instigate the need but often with the wind came rain, hail, snow and even lightning. All of which put a challenge to the job and a bit of living dangerously.

Then came the deluxe method of antenna alignment – the motorized turner that operated from a box on the top of the television. Even that had its own personality with the ever consistent "ker-thunk, pause, ker-thunk, pause, ker-thunk, pause" as the antenna ker-thunked into position.

Today's youth are masters of the remote control that manages hundreds of channels on cable or satellite. Not only do they remain clueless about the character-building efforts of antenna management, they have no idea about the curious wonders of the "sign-off."

You remember the one where, as children, we stared at a screen with a target looking emblem accompanied by a piercing ringing sound and wondered what would happen next.

Yes, Virginia, there was a time when television was not available 24/7.

A Generation of Yesterday's Cowboys

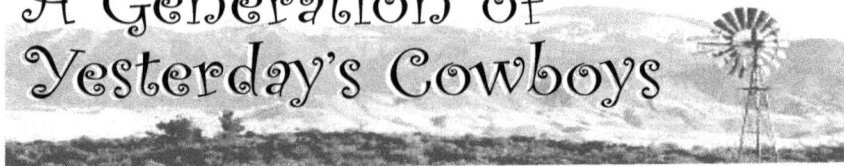

An interview I did last week got me to thinking about the generation of cowboys that went before me. Things were very different then...then being the decades of the 40s, 50s and 60s.

Cowboys in those days weren't always born cowboys. Cowboys that fathered generations of today's cowboys were often "made" not born. The freedom and excitement of life on the range was very alluring to many young men whose parents intended for them to be doctors and lawyers or even just hoped they'd get a real job.

My Dad and uncles were some of those 30s babies that grew up to be fine cowboys and respected cattlemen. Their on-job-training started very young and was a crash course in the finer etiquettes of a ranch cowboy.

My Dad died before his stories got set on paper so what I have of him are memories and what he taught me. I put the burden of recording those days on paper on his younger brother.

My uncle wrote: "At 1947 at 13 years of age I got my first job at trying to be a cowboy working for the KC Ranch in Colorado. They had three ranches. One at Gardner, Colo., one in the Davis Mountains west of Pecos, Texas and one in an isolated area about 70 southwest of Carlsbad, New Mexico, 70 miles northeast of Vanhorn, Texas and 70 miles northwest of Pecos."

"The summer I was 14, I went to the Davis Mountain Ranch to work. We spent the summer doctoring cattle for screwworms. The summer rains started and it would rain between ½" and 1 ½" every afternoon. Excellent weather for screwworms."

"I went back to the KC north ranch in West Texas the summer I was 15. We'd move or work cattle from daylight until about ten and then spend the afternoons shoeing horses or doing whatever else needed to be done."

"The summer I was 17, and had just graduated, I was going to win my fame and fortune rodeoing but I needed a nest egg to get started. So back to the KC Ranch in West Texas I went and let it be known I was going to have a career at riding bucking horses."

"I told them I was only staying long enough to get myself a nest egg. When the cow boss cut my string of saddle horses he cut them with my 'future career' in mind. Every one of them would buck quite a bit every time you would ride them."

"I had one horse named Half Dollar. He had bucked off about everyone who had ever had him in their string. We had a herd thrown together and all of a sudden he started to buck and bucked for a good little bit. I was lucky and rode him. Everyone gathered around me and was talking about what a bronc ride that I had put on. About that time Half Dollar shook real hard and I fell off – right flat on my back."

"When I left to pursue my rodeo career, everyone wished me luck and told me to not let any horses shake with me."

Along with the usual job skills, every cowboy inevitably learns a little humility.

Fly Away on a Haystack

As you drive down the highways across America, most rural areas have something in common that is taken for granted as part of rural life. Hay stacks. Coast to coast, border to border they rise up representative of the fruit of the land.

I grew up with hay meadows as much part of ranch life as were the cattle, horses, chickens and pigs. With the brome and timothy grasses as tall as I was, the world beneath the tops was a maze of sweet smelling grass and places to hide. Wearing out the knees on my jeans, I spent hours crawling through the grass hiding from my brothers who thought the game was delightful fun.

One of those firsts in life happened when my grandfather let me drive the tractor pulling the baler. I felt so grown up. Baling hay when I could barely reach the pedals on the tractor! But the best part was to come when those bales were placed in long and high stacks in the hay lot.

To my dad, those stacks represented a long winter of feeding cattle in snow filled pastures on cold windy days. To his children, they were a play land that allowed imaginations to create worlds they had never experienced.

Those haystacks became submarines, airplanes, sailing ships, sky-scrapers, wagon trains and every now then, just haystacks.

My brothers and I would create a story to go with the item we'd made of the stack for that day. We flew to parts of the world we couldn't even spell. We had offices in a high rise building even if it was primitive, with a rope instead of an elevator.

We sailed seas and fought pirates. We went west with Major Adams and Flint McCullough and circled the wagons under Indian attack. The daredevil in the boys would come out when the ropes were strung across the lot to other stacks. Making that trip stack to stack hanging from the rope was the same making the crossing rim to rim of the Grand Canyon.

And then there was the one game that we older two are most proud of – pulling out a few bales to make a "pit" and leaving our younger brother there so we could run off to the creek to go fishing without the little pest tagging along.

We had no idea at that time what the world would hold for us. When we got home from school we would run down the hill, climb over the pole fence to that baled wonderland and leave any thoughts of the future behind. Life was so simple and so innocent. Well except the part about hiding from our brother.

Today, the four of us that lived that simple childhood life smelling of fresh cut hay just in from the field, have scattered to a world of corporations, traffic and neighbors next door.

Yet for us all, the smell of an alfalfa field in bloom or the sight of a haystack standing tall brings instant recall of those times.

I don't know what part of my character was formed on those haystacks but I'll always believe it played a big part of who I am today. And that pest of a brother? He still is.

Back to School...
Now and Then

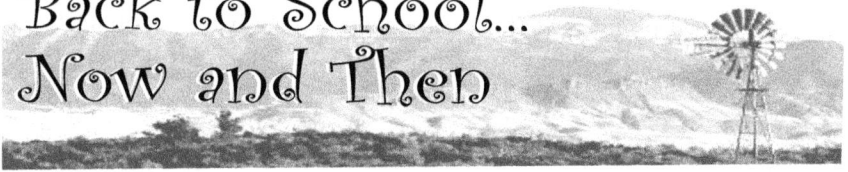

The kiddy song goes "The wheels on the bus go round and round….. all through the town." Once again it is time for the big yellow vehicles to be put back in gear to carry children to the school houses for another season.

It is a decades-old routine that often brings with it a wave of nostalgia for many, me included. There are at least three and sometimes four generations of America's students that can still recall how much different things were "back then."

I started school in a one-room country school house along with less than a dozen others in grades one through eight. I began in the first grade with three other children because they either hadn't invented kindergarten yet or it hadn't reached the rural mountain regions of Southern Colorado.

We shared one teacher, a huge wood stove in the middle of the room and a chalk board that was filled with everything from primer words to eighth grade math problems. And yes, Dick, Jane and "go Spot go" were part of my formative years.

We all brought sack lunches as there was no lunch program – free or otherwise. We had a "Boys" and a "Girls" bathroom option in the form of wood outhouses at the back of the school yard. To this day I don't recall if they were one- or two-holers.

Recess offered baseball, a set of swings with board seats complete with splinters and the usual playground games that required no equipment, only imagination. The apple tree at the back of the school yard carried the legend that it had begun when students from earlier years had thrown their lunch apple cores in a pile in that spot. I thought that was a magical story.

We had a Christmas play on a small stage that became available by removing part of a wall between the classroom and the backroom of the school house. I can remember being terrified to stand alone and sing my part of Jolly Old St. Nicholas. Now that I am aware of my lack of musical ability, I know there were many reasons to be afraid.

Gasoline was less than 30 cents a gallon but the drive to get me from the ranch to the school was often country dirt road difficult. I spent many weeks in the winter bunking with my teacher and her family on a ranch closer to the school.

While that saved on long walks when stuck in the snow, fuel costs and wear and tear on vehicles, it often made a little girl very homesick. But I loved my teacher and still do today. She reads these columns faithfully every week.

She did what teachers are supposed to do. She sparked in me a desire to learn and the belief I could do anything I set my mind to do. Even if she did put my hair in rag curls.

The world of 1958 was in transition. That unique one-room school experience lasted only a year and the tiny school at Malachite, Colorado was "consolidated" and the students bussed to nearby Gardner.

I thought I'd hit the big time. There were at least eight kids in my class. Better yet there was a filling station across the road that sold penny candy at lunch time. Things were looking up.

The Malachite school was built of rock – a foundation that formed walls up to large windows and a peaked roof that held a bell tower pointing to the blue Colorado sky above.

I'd like to believe that one important year there began to form my life in the same way – rock solid underneath and always reaching to the sky.

Thank You for Calling Me Cowboy

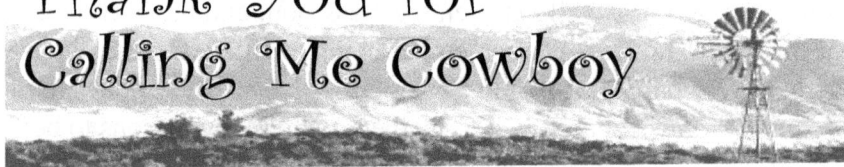

You wouldn't think it wouldn't happen here in rural small town New Mexico, a state that is home to cowboys that are famous, infamous and some just regular real cowboys.

But it did happen. My 10-year-old son came home from school totally insulted because his classmates told him he was "too cowboy." Disgusted he said, "They don't like me because I'm a cowboy."

I tried not to laugh. I knew he was serious and didn't know how to take what he perceived as insults. I told him what they had said was not really the insult they wanted it to be. Without realizing it they were giving him a high compliment.

He looked puzzled so I continued to explain to him just what being a cowboy meant.

"Cowboy" is not just a job punching cows, breaking horses and roping critters. It is not just a cowboy hat and a redneck attitude although there is nothing wrong with either.

It is an attitude of integrity and a heritage. It is way of living and doing things. It is manners, responsibility and living ethics that the world as a whole has long forgotten.

The Code of the West says to ask no more and give no less than honesty, courage, loyalty, generosity, and fairness.

Cowboys thrive on simplicity. Simplicity is something cowboys just naturally understand…low stress, a relaxed demeanor and a basic and simple lifestyle.

Listen to the practical wisdom in these nuggets of cowboy logic:

"The only way to drive cattle fast – is slowly."
"No matter where you ride to – that's where you are."
"Never approach a bull from the front, a horse from the rear or a fool from any direction."

Loyalty runs deep in cowboys. "Ridin' for the brand" isn't just working on the outfit that pays you. It is a deeply imbedded strain of integrity that requires doing things to the best of your of ability on whatever task is set before you – whether you like the job or not.

Loyalty to his employer, loyalty to friends, and loyalty to God, country and family all make up the man we call cowboy.

In *Lonesome Dove*, when Woodrow Call promised his friend Gus McCrae that he'd take him from Montana back to Texas to bury him when he died, he meant it. Yes, it was a movie script, but it so vividly portrayed the way things are between men like that. I know men today that would do the same no matter the odds.

Honesty, integrity and standing for what is right are traits that a cowboy wears always. Quoting from *Cowboy Logic*, "Honesty is not something you flirt with – you should be married to it!"

These are people who work hard all day and go to bed tired. To them integrity and honesty are words that still have meaning.

To a cowboy (today and in the old West) – honesty is and was a given. A man's word was his bond and a handshake deal was a done deal. Honesty seemed to simply come as a second nature. And for a cowboy, the difference between right and wrong has no gray area.

I reminded my son he came from a long line of cowboys and that was a good thing. I told him the next time he was "accused" of being a cowboy he should simply say, "Thank you. I'm proud to be one."

Only in a Small Town

Smalltown America: an icon of Americana that stirs memories of simpler lives, simpler times.

Many of us are still blessed to live near or in one of those little towns. More still have memories of "that's where they came from originally." People that are born, raised and live in metropolis areas don't know what they have missed!

Everyone waves at everyone and it's not an obscene gesture. The wave doesn't need to mean you know who you are waving at. It is just the friendly nature of the local folks. They wave in town and they wave on the highway leaving town.

The local bank president moonlights as a ranch hand on weekends when his family works cattle. He runs a branding iron or drags calves to the fire right along with the hands that do it for a living. He's a third generation rancher that ended up with a suit-and-tie job.

Signing a ticket in most towns means the Visa or MasterCard part of the sale. In small towns, it's a charge ticket at the hardware store and requires only stating, "this is for old so and so and they told me to put it on a ticket for them."

Getting directions is often a challenge in asking the right questions. Most locals will answer in terms requiring knowledge and a memory of at least the last 25 years.

"You make a turn out there by the old Smith House and then go down that road until you have to turn that corner where those kids wrecked their car that time, then it's not far past that." Using "the stop light" as a reference point is not uncommon.

The UPS man knows your phone number by heart and is able to follow the instructions mentioned in the last paragraph. He will call you and tell you he left a package for you at the hardware store and who it was from in case it is something you urgently need.

I find this a unique service in that UPS won't come out to the ranch any more and it lets me be part of a modern civilization in spite of it.

You know who is in town by whose cars are parked at the post office and the courthouse. If you don't run into them there, they'll be at lunch in one of the few eating spots in town. You never eat out without having to catch a visit with someone you haven't seen in awhile.

Most of the vehicles on the street still have their keys in them. The doors will be open and you can leave your neighbor that dreaded sack of "extra" zucchini he's been dodging for a weeks.

The manager of the grocery store can be seen pushing a cart down the aisle filling an order for someone that called it in. If something you need isn't on the shelf by brand name or kind, tell him and he'll get it in the next shipment.

FFA and 4-H are a big part of most kids' lives and everyone in town knows about it. The graduating class is small enough to fit in two Suburbans for a trip. Homecoming is a big deal and school sports are the center of most of the winter conversations in town.

Merchants know you, know your children, and know where to find you if your children are looking for you. The police station in town is not a scary place since the chief has school children; he's a familiar face to the kids.

Small-town America. Where the kids can't wait to grow up and get away.

Then they spend most their lives wishing they were back in that simple life among people who really care about each other. It's that caring familiar attitude that draws those leaving cities to come among us.

The Day the Willows Gave Me Away

The topic of most embarrassing moments is not an uncommon one in print or in conversations.

For some reason people who call themselves your friends love for you to relive any moment in time that made you feel really uneasy, awkward or with that timeless wish to be able to crawl in a hole.

First, let me say it has occurred to me often that perhaps I need new friends. A few of them actually started and participated in a poll to vote whether they thought I'd ever been embarrassed or not. Somehow, my image has taken on a cast iron sheen that makes some think I'm tougher than I really am.

Of course, out of 21 votes only one person who yet remains anonymous voted "yes," they thought I'd been embarrassed. Who needs 21 friends anyway except perhaps someone running for a political office.

The question right after the "If..." then becomes, "Why or what happened?" They wait like rabid dogs for the details of the moment that goes down in history to haunt you. So today I'll set the record straight.

I was about 8 years old when the first very embarrassing moment in my life happened.

A friend and I had gone riding in the mountain meadows on the ranch in Colorado where I grew up. It was a hot summer day and the beaver ponds in the creek looked invitingly cool. The idea occurred to us both about the same time.

We tied up our horses, stripped down to our underwear, waded in chest deep and splashed away until it was time to go home.

Riding horses with wet underwear under your jeans is a chafing idea. So I decided to abandon the wet item by tossing them deep into the willows where I was sure they wouldn't be found in my lifetime. I donned my jeans and we rode on home.

Not two days later, my dad, with me along, took a group of guests trout fishing along that same creek.

Fate was not kind to me nor was his overactive ornery sense of humor. He spotted the discarded undies almost immediately, poked them out of the willows with the end of his fishing pole, and paraded them in public like a flag waving in the breeze.

"Are these yours Julie?" he asked knowing darn well they were. He was grinning from ear to ear at his discovery and my embarrassment. You just didn't show folks your undies anytime and this was just over the line!

I wavered between extreme embarrassment and fear of what was to come next when he remembered he'd specifically told us not to go swimming without permission or supervision.

Fortunately for me I didn't die of embarrassment although it seemed to be possible at the time. My dad only scolded me that time in lieu of tanning my backside for disobeying and I've often wondered if I got off light because he so enjoyed the moment. No one in my family has EVER forgotten the event and it still comes up from time to time at family gatherings.

Was that the only time or last time for the blood to rush to my face in sheer embarrassment? Well of course not, but it only takes once to prove the poll wrong.

Those Things We Remember of Grandma

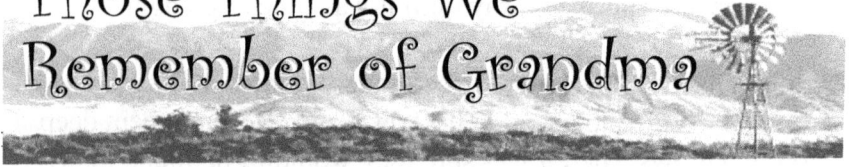

The author of an essay called *Grandma's Apron* unfortunately seems to be unknown. The missive has been traveling the information highway and has landed in my email no less than five times.

The poignant recall is simple enough to touch everyone that ever had a grandmother even if she didn't wear an apron. It reads:

The principle use of Grandma's apron was to protect the dress underneath, but along with that, it served as a holder for removing hot pans from the oven; it was wonderful for drying children's tears, and on occasion was even used for cleaning out dirty ears.

From the chicken-coop the apron was used for carrying eggs, fussy chicks, and sometimes half-hatched eggs to be finished in the warming oven.

When company came, those old aprons were ideal hiding places for shy kids; and when the weather was cold, grandma wrapped it around her arms.

Those big old aprons wiped many a perspiring brow, bent over the hot wood stove. Chips and kindling-wood were brought into the kitchen in that apron.

From the garden it carried all sorts of vegetables. After the peas had been shelled it carried out the hulls. In the fall it was used to bring in apples that had fallen from the trees.

When unexpected company drove up the road, it was surprising how much furniture that old apron could dust in a matter of seconds.

When dinner was ready, Grandma walked out on the porch and waved her apron, and the men knew it was time to come in from the fields for dinner.

It will be a long time before anyone invents something that will replace that old-time apron that served so many purposes.

There are many things that bring an instant recall of my grandmother.

Even at the ranch, she wore a dress everyday – well into the 1960s. She seemed old to me when she was in her 40s, but looking back I now know hard work will do that to a person.

In the winter in those cold old ranch houses, the kitchen was kept warm with something she was cooking and the cooking went on most all day.

She would open the oven door to help with the warming. Early mornings you would see her backed up to the oven to warm her back side. She would lift her skirt to let the warm air flow under. That visual will stay with me like the apron did with others.

She cooked, canned, baked, gardened, sewed and sang the whole time she worked. "Send me the pillow that you dream on," she would melodiously belt out like she really could sing. "So darling, I can dream on it too."

The world today is pure convenience and automation. I often wonder what she would think of all the advantages we have.

The invention of the crock pot that slow cooks its way to a meal ready-to-eat when we arrive home after a day at work has been rated right up there with the invention of the wheel.

Not long ago when I was grocery shopping I spotted what I first thought was the absolute height of laziness in conveniences: a packaged crock pot dinner product all cut, diced, seasoned and ready to drop in the crock pot to slow cook all day.

Again I thought, "Pure laziness." So I bought only one.

Willow Switch Mentality

Most kids are pretty easy to entertain and I believe country kids are even easier. Pick a generation and they all have a long list of things in common. It was an era of simple entertainment by making do with what was available.

Until the age of satellite dishes, remote rural children knew very little of television. I grew up in the time when we got one channel. If the wind was blowing, one of us four kids was climbing on the roof periodically to turn the TV antenna back where it would pick up the signal. Saturday morning was the only cartoon opportunity and *Bonanza* was a Sunday night ritual.

We fished and played cowboys, Indians and hillbillies (we didn't know we were one) in the willows along the creek. We camped at creek side, ate burned marshmallows and told ghost stories to scare ourselves all the while able to see the house lights from our camp.

We rode our horses anywhere we wanted to go, just as long as we told Mom where we were going and when to expect us back. Often we packed a lunch and would be gone all day on some adventure into the hills. My mother had to have nerves of cast iron now that I think back.

In the winter we shoveled the snow off the beaver ponds to ice skate on. We let the barbwire down on a fence so we could have a nice long toboggan run without having to duck under the wire. Before the invention of saucer sleds we used the round saucer-like top off the grainery to fly over the snow.

We didn't have a sleigh or a horse that would pull it but we did have a Shetland pony and a 6 foot toboggan. Now the pony was not broke to drive or pull and his favorite thing was to run off and head back to the barn. Not to be daunted by details, we used every rope and loose strap from the saddle shed to fashion a harness to hookup the toboggan.

Then we'd lead ole Mickey Mouse – that was his name – to the top side of a long hay meadow, jump on the sled and let him do his run back to the barn trick. The ride was wild and rarely did any of us make it back with the horse and sled.

Spring brought the snow melt run off and the creeks would run high and bitter cold. It was more than a kid could stand not to get in that water and do some wading when ole man winter finally left the high country. My dad would scold and threaten us to stay out of the high running water.

But as kids will do, we'd get in just as soon as we thought he wasn't looking. And just as soon as we stepped foot in the water he'd come with a fresh cut willow switch to wear out on our back sides all the way to the house.

I think it would solve a lot of the world's problems if we had more of that life with one TV channel, more Mickey Mouse-pony kind of fun and a few more well used willow switches.

When I mentioned to my son he needed a willow switch across his bottom he asked "What's a willow switch?" I had to cut a juniper branch and show him.

You Know You are a County Fair Mom When...

The glue that holds together every family participating in a county fair is typically and unequivocally the person known as "mom."

During the days prior to the fair, Mom is charged with a long list of assorted responsibilities that must be completed before the family ever pulls out of the driveway headed to the fair grounds.

Usually at the top of that list are food, shelter and clothing. It will be she who gets the camp trailer ready for a week's invasion. From blankets to baloney, it is Mom who must prepare the mobile castle with all the comforts of home.

That of course requires endless planning lists of shopping, cooking, and packing. Buy it, bring it home, cook it up, load it in the trailer and then try to be kind when the family insists on eating three squares at the concession stand.

On most occasions Mom is also an additional stable hand. She will be washing pigs while the child is walking goats. Dad will be clipping lambs while Mom is making a mad dash to the grocery store for forgotten ingredients needed for the contest cake to be baked the very same day the pigs are due in the show ring.

As kinfolk and town folk wander into the show barn, it is not hard to tell the "fair moms" from the visitors.

You know you are county fair mom when:

□ For two weeks before the fair tempers flare and everyone is sniping at each other but mostly at Mom. After all, it's her fault they have to get things done on time.

- Every night you talk about pigs in your sleep. Once at the fair, you are so exhausted, but can't sleep or even dream about pigs.

- For three days you have washed and ironed a weeks worth of clothes for the entire family. The family arrives looking starched and pristine while you are at the fair with no make up, wearing the same clothes you have had on for two days and flat hair.

- You finally arrive at the fair grounds intact with the entire family, four pigs, two lambs, a goat and a camp trailer only to realize two of your daughter's novice 4-H projects didn't make the trip. And it's too late to go back and get them.

- You find yourself working up a sweat and getting no respect for the hard work it is to give orders all day to a family that is never listening. Others will recognize you as the woman at the edge of the show ring holding a halter, bucket, brush, blow dryer, clippers, spray bottle and can of Show Sheen.

- When you don't care who sees you grab your child by the ear and drag him across the fairgrounds while giving him a tongue lashing. Any one foolish enough to comment gets the "Mom look" that says "You want some of this?"

- You are always the one who last had whatever it is they can't find. Four days into the fair and couples are rarely seen speaking to each other and then only with a hint of civility. This wonderful family event has about broken the need for togetherness for a season.

- Your schedule for buying new footwear for the entire family is always after the fair knowing that any shoe that comes to the fair will be full of pig stuff, mud and carry an eternal odor.

- By the end of the week when the family loads up to go home, no one is speaking to anyone. But everyone will weakly wave and say "See ya next year!"

Coming Home Again

I am the favorite daughter and the favorite sister. My brothers will have to agree with both statements because I am the only girl in our family of four children.

I am also the eldest of the pack of mountain children, which sounds better than hillbillies, but really, that's what we were.

This weekend we will all be under the same roof for the first time in a dozen years.

I'm not sure how that happens; how a family once so close can get so scattered and caught up in life that we forget to come home, but it has.

My mother is ecstatic about the gathering; as well she should and would be.

As all mothers will do, she will feed us well, give us sage advice and tell us stories from the old days we've probably heard multiple times.

And best of all, she will scold us like she did when we were all under the age of 12. There is something comforting about that.

The teasing and the bantering will be non-stop. The grandchildren will hear tales about their parents that will amaze and delight them. They are old enough now to have an adult concept of their parents as children.

They also begin to understand why their parents are smart enough to know what they are up to; that parents really aren't clairvoyant, just experienced.

We will once again prove that tattling to your parents is not something that stops with adulthood.

People who are now getting AARP propaganda in the mail will still find delight in reminding their mother just how bad their brother or sister was as a teen.

"Mom did you ever know that Julie was really not at her friend's house like she said she was? You knew about that party, right?"

"Mom, did Lonnie ever tell you about the time .... ?" This conversation could take half the weekend.

"Mom, you do know that Bruce wasn't really sick because of your fried chicken, right? Fried chicken doesn't give you a hangover."

"Mom, we know Jim is your favorite but only because he's the baby. He's not really any better than we are. He just got away with it more often."

Sibling relationships slide right into adulthood without much adjustment.

I'm still the bossy older sister with little tolerance for their nonsense and have lost none of my ability to tell them so. They have lost none of their ability to ignore me.

The three "boys" will size each other up for what life has dealt them in the way of careers, wives and children.

After that short assessment is over, they'll all hit the cookie jar looking for mom's specialty – chocolate chip cookies.

The visit probably won't be long enough, but maybe it will fuel the desire to do it again before a decade passes by.

We aren't special or unique. Nor are we any less or any more dysfunctional than most families.

What we are is family. And for that, we do know we are blessed. Even if we don't always show it.

Rural
Holidays

Christmas Shopping Guide for the Cowboy on Your List

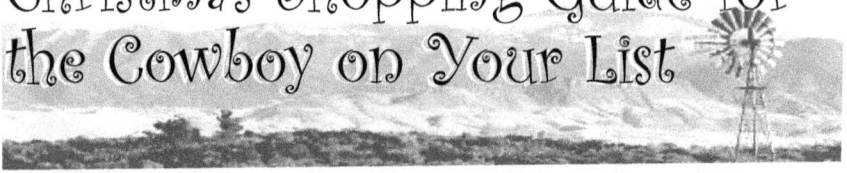

Let's say you have a favorite cowboy you want to buy a gift for this Christmas and let's say you are new to cowboy shopping. Let me give you an extensive list of cowboy shopping do's and don'ts.

Keep in mind there are always exceptions to the rule, here are this year's DO NOT BUY:

1. Anything made of polyester but certainly not a polyester sport coat.
2. Designer socks or silk jockey shorts.
3. Tofu anything and white wine.
4. A sweater vest.
5. A salad spinner.
6. A George Foreman get all the fat off the meat grill.
7. A vehicle without 4-wheel drive and less than ¾ ton capacity.
8. Season tickets to the opera including opera glasses.
9. Tennis, backgammon, or croquet lessons.
10. Complete set of Danielle Steele romance novels.
11. Driving gloves.
12. Gold chains and earrings.
13. Velour, embroidered or lace trimmed articles of clothing.
14. Self-help books on how to get in touch with your feminine side.
15. Speedos.
16. Any beverage holder with less than a four finger handle.
17. Beano – he won't use it.
18. A manicure set with anything smaller than #9 wire pliers and a hoof rasp.
19. Line dance lessons.
20. A day at the spa.
21. Subscription to *GQ* magazine.
22. A gourmet cookbook.
23. Color coordinated shirt and tie by Ralph Lauren.
24. A smoking jacket.
25. A set of instructions for anything.

Sure-fire pleasers under the Christmas tree TO BUY the cowboy are:
1. Anything from the feed store.
2. Anything from the hardware store.
3. Tickets to the National Finals Rodeo and a subscription to *ProRodeo News*.
4. Membership to his favorite rodeo association and a gift certificate for a years worth of entry fees in his chosen event.
5. Make that vest a leather, canvas duck or nylon down-filled one.
6. Anything labeled Wrangler, Levi, Stetson, Tony Lama, Justin, or Carhart.
7. Cast iron – especially if it's a skillet complete with the promise of a years supply of fried steak, potatoes, okra, bacon, eggs, and even refried beans.
8. A one-ton flatbed truck complete with a propane fuel tank, grill guard, headache rack, gun rack, mud and snow tires all the way around and a new chain, shovel and axe in a tool box in the back.
9. Tools that say "life time warranty, guaranteed forever."
10. A good pocket knife, made in the USA and a new whetstone.
11. Lots of cammo and ammo.
12. Pinto beans by the gunny sack full.
13. Cookbook called 101 ways to cook venison.
14. A book called "Teach your woman to run a trap line."
15. New five-buckle overshoes – boot style.
16. Heavy duty one-gazillion candle power spotlight for calving season.
17. A gift basket full of beanie weenies, spam, Vienna sausages, beef jerky and huntin' license good for anywhere to shoot anything.
18. A roll of Copenhagen or Skoal for his stocking hung by the fire with care.
19. A tooled leather, belt mounted cell phone case.

These ideas of course are only suggested as guidelines and can be mixed and matched to suit the cowboy in your life. Happy shopping!

Gift Shopping for the Rancher's Wife

The season started in the retail world right after the garden supplies were moved to the back room and school had not yet started. But that was just a warm up to what explodes cash register drawers the day after Thanksgiving – the Christmas shopping frenzy.

As is for most things, gift shopping at the ranch is a pretty laid back procedure. I'm not saying a lot of thought is not put into choosing the perfect gift, but "perfect" is subject to interpretation and you can almost always factor in functional and fundamental.

The gift that wins the tally of given most often from him to her is an axe. I know that will shock many of you that don't live down dirt roads, but an axe is essential to the time of year the gift is given – Christmas and the middle of winter.

The axes have come single bit, double bit and often tied with a red bow the size of a pick up truck in an attempt to make it festively palatable. Some have come with a flashlight as an extra gift so wood or ice could be chopped in the dark. Often a note is attached saying, "I promise to keep this sharp for you."

Another common feature for the ranch "him to her" gifts is the "who really wants or needs this?" Gloves that are too big for her and fit him perfectly are regular offerings under the Christmas tree as are new saddles when she rarely rides, horses she never will ride and that absolutely stunning truck tool box that unfortunately won't fit her SUV. A complete assortment of hand and power tools also fit into this category.

Never say the gifts aren't truly appreciated. One wife I know got a new cattle guard. It was to be placed where she had to open and close a gate 15 times a day coming and going. She would not have been happier if she had gotten big blue diamonds.

Always thinking of the Little Woman's health and safety as well as her viability as the best if only help he has, he will gift her with things to keep her warm and useful. That list will include insulated coveralls, down-filled everything including lingerie, and even a new rifle to carry on her 4-wheeler to shoot coyotes while she is checking heifers and new baby calves.

Buckets of all sorts rate right up at the top in frequency of gift types – feed buckets, milk buckets and buckets to bail drinking water from a well or cistern. One gal was so proud of her new international mop bucket with the "Caution" warning in both English and Spanish. It had wheels and every feature you could imagine except a back-up alarm.

Other gifts have come with the possibility he is going to get shot if her sense of humor isn't at its peak. An oversized personalized "fire flapper" was indeed given to a wife I know with the justification that "she's a big girl so she may as well do some good when she is beating out a grass fire."

Feed stores, hardware outlets, saddle and cowboy tack dealers as well as livestock sale barns across rural America are standing by to serve the rancher in this season of giving.

Gathering Around the Old Oak Table

The round oak table in my mother's dining room is as much part of our family history as our family names and all our relatives.

No one knows exactly how old the table is, but speculation with the dates, we do know it puts it in the 70-80 year old range.

It was left behind in an abandoned homestead in Colorado. It was gathering dust in a shed and had been used for a butcher table – complete with saw cuts all around.

In 1956 my mother and dad brought it home. They sanded it down and refinished it for the first of three times in its life with our family. One by one the saw cuts were sanded out of the oak except for those too deep to remove.

Using money earned from cutting and selling Christmas trees, they spent eleven dollars on raw oak boards to make five additional leaves for the table.

Dad had no power tools to work with so every step of the way was by hand. Each leaf has a number penciled on the back so it is placed in the table in the correct order to make the pegs fit in the holes properly.

In an era when a dollar was a huge sum, they turned down a $500 offer for the finished product. The natural quarter-sawn oak table had value to the world but never more than it did to us.

My family has lived around that table. Always extended, with at least two leaves, it can easily seat eight, and fully extended it let us seat 20 or more during the holidays.

It was those times as a child I thought life was the very best. Never enough chairs, the piano bench would seat two kids and the flour barrel one more. The "little" kids had to sit at a card table. In a rite of passage of sorts, it was an honor to dine with the adults even if you had to sit on a flour barrel.

I remember the holidays as always noisy, fun and with lots of food lined up on that oak table. I can still hear the singing in the kitchen when my aunts and grandmothers and mom were doing the dishes and putting away the food after the dinner. Nobody could sing very well, but nobody cared.

If it could tell its story, the table would tell you how we have laughed, how we cried, how we celebrated and how we mourned for these near fifty years – all around that round oak table.

It would explain the small dent that was made when my mother pounded the pearl snaps on the Western shirts she made for my dad. It would tell of the many late nights of family card games, Monopoly, and Parcheesi accompanied by gallons of Kool-Aid and bowls of popcorn. It would tell you of the frosting for the hundreds of Christmas cookies and the egg dye for as many Easters.

Looking back I think the oak table is a lot like life. It has seen many seasons, many events and holidays, many decades of living. It has suffered cuts and bruises, been relocated, rearranged and then refurbished. Its usefulness was never a question.

This holiday season my family will again gather to celebrate. There is now a fourth generation in our family that is learning about life around the old oak table.

Holidays Steeped in Family Values

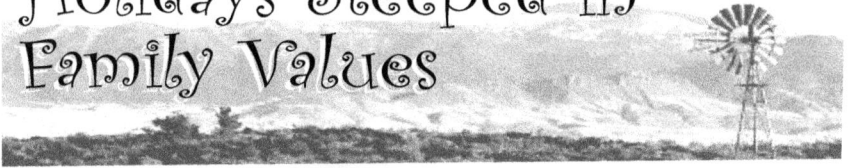

Pumpkins have taken on a less toothy look in recent weeks and have been joined by turkeys, pilgrims holding platters of food and cornucopias spilling over with vegetable bounty.

If there is any doubt which season is headed your way, the commercialism of it will quickly bring it to your recollection. It was not too soon to find Santa in August immediately after the stores cleared out the garden supplies.

As a parent, I have always swum upstream through the commercial holiday onslaught trying to find and keep the true meaning of the days we celebrate in November and December.

While always giving reverence to Christmas and honoring the birth of Christ, the holiday for me has solidified the tradition and importance of family. The values for both were what I strived for my children to experience, knowing the world would attempt to woo them to a shallow version without sanctity of meaning.

My childhood holidays were grounded in family traditions in a very rural sort of way. Holiday dinners involved visitation from 30 or so of our closest relatives; not always the ones we liked the best but family just the same. Most of them drove a long distance on rough roads to spend the day with us at the ranch.

We would gather around my mother's round, but for the occasion elongated, oak table filled with tons of food, dishes that didn't match and never enough chairs for everyone. No one cared. Ambiance had not been invented yet in our part of the world.

As soon as the blessing was said the noise level picked up several decibels and the platters and bowls began to circle by the hungry kinfolk. Coming-of-age meant you were allowed to sit at the big table with the adults and not a card table with the smaller kids that were spitting mashed potatoes, spilling Kool-Aid and sticking vegetables up their noses.

Thanksgiving officially opened the holiday season that continued through the month ahead with traditional events at home at the ranch joined by assorted branches of the family.

My brother's birthday was in early December. Not to be slighted because of the time of year, the traditional family birthday gathering would convene, gifts offered and another family meal served around the enduring oak table.

My dad's birthday was Dec. 18. That day was, always and traditionally, celebrated by the return of the relatives. They would trudge through hip deep snow, cut a Christmas tree from a steep hillside and drag it back to the pickup while children ran, played, tobogganed and initiated snowball fights amidst the adult dedication for a perfectly shaped tree.

The day would close in celebration of the birthday with yet another family dinner. If the kinfolk lingered long enough, they were entertained by watching my mother try to decorate a 10-foot spruce tree with four children, hyped with Christmas excitement, forcing their help upon her. Always on that day, no matter the circumstances, the tree my dad cut was transformed into the most beautiful Christmas tree in the world.

It twinkled with lights that should have burned it down but we didn't know that so it didn't. Strands of glass beads draped the branches adding symmetry in a festive way. The sparkling glass balls added color and dimension and the shining strands of tinsel, tediously placed one piece at a time, reflected the beauty in a shimmer of icy reflection.

The tradition of decorating on Dec. 18 is now in its third generation. No matter where I am, no matter the climate or the zip code, those memories represent to my very soul the foundation of family. I didn't know what was happening at the time, but I never lost it.

As Thanksgiving beckons us to reflect on those things that beg for a grateful heart, I begin the praise with my eternal thanks for first, the love of God, and second, the love of a family that has transcended a lifetime.

If I have nothing else to give to my children, there is that.

Hearts, Flowers and Chicken Feathers

Saddle up boys, here it arrives every year, ready or not. Valentine's is February's claim to fame. Commercialism has painted the world with pink and red hearts and accented it with chocolate.

Some will attest to the theory that Valentine's Day was invented as a clever ploy to stimulate the economy in an otherwise financially sluggish time of year. Greeting card companies, florists, jewelers and chocolate manufacturers who flourish because of the promotion would have to agree.

Valentine's Day advertisements, even locally in a rural part of the world, promise evenings of lasting romance and adoration if you will just come dine with them for only $175 a couple. I don't believe too many pickup trucks will be leaving the ranch for that offer.

But there will be some "romantic" gestures made out on the range. It may not be wine and roses but a cowboy on a Valentine's Day date will offer his heart's delight a romantic late night walk through the frosty pastures for a "just once more" check of the heifers. After all, it is calving season.

I know a gal who books her husband and herself into the dentist for a teeth cleaning every Valentine's Day. "Nothing says 'I love you' like a plaque-less kiss," she claims.

But this year's top story demonstrating "true love" in the best way they know how comes from a local ranch. The boss left home in the morning, as usual, to go make his rounds feeding cattle and checking waters.

Millie, the Border collie, was pleased he was going alone because that meant she got to sit up front and ride shotgun in the feed pickup.

Quite a distance into the feed route and miles from the house, the rancher happened to catch a glimpse of something in his rear view mirror. He stopped the pickup and walked to the back only to find one of his wife's beloved and treasured chickens riding on the flat bed of pickup.

At this point in time, this man had many options before him, none of which would have been healthy for the chicken.

Most men would have, at the very minimum, denied all knowledge of ever seeing the hen and more than likely left her in the pasture to fend for herself against the natural order of the food chain in the wild. Chickens usually rank pretty low on the compassion scale for cowboys.

But knowing how much his wife adored her birds of all kinds and especially her hand-raised chickens, he gathered the hen up and put her in the front of the pickup on the seat between Millie and himself.

Millie was indignant and completely insulted. She turned her head, nose in the air, and stared out the passenger window the remainder of the trip trying to her best to pretend there was *not* a chicken in the seat next to her.

The rancher finished his feed route and returned home a few hours later, the hen nestled tight against him for warmth.

The sight had to be one of those rare moments none of us actually ever see. The visual of this guy driving down the road with his dog and his wife's chicken in the front seat of the pickup is enough to put anyone into fits of laughter.

It also makes a good "true love" story. Not many, chicken lovers or not, will miss the depth of the affection it took for this guy to cozy up to a chicken, even for the Little Woman.

The Christmas Pony

My dad hated ponies, Shetland or otherwise. His heartfelt belief was that if you wanted to ride – ride a real horse.

No one seems to really know what possessed him to bring home a Shetland pony for his kids that Christmas. We owned plenty of "real" horses.

Somehow in a horse trade he ended up with this short, barrel round, pitch black, Shetland. He was trading off a perfectly good bay two-year-old real horse for some cash. Somewhere in the deal, this small want-to-be horse got hauled back to the ranch.

He named him Mickey Mouse. Not for his color and not for the Disney character but because this midget was a mickey mouse version of what a horse was supposed to be.

Mickey repeatedly proved my dad's theory on why not to own a Shetland pony and it began on Christmas morning.

At daylight, Dad went to the barn to do chores. He had the pony hidden in the barn but Mickey needed water. We had no water hydrants or tanks in the barn or the corrals. Water was the creek that ran along the bottom of the small trap below the corrals.

So Dad bridled the wee equine and jumped on him bareback to ride him to water. At some point during the process Mickey reared up, sliding Dad off his back. He landed hard on his back pockets on the hard frozen ground, breaking his tailbone.

Mickey Mouse defined in every way spoiled, barn soured, obnoxious and aggravating. If you rode him anywhere, he spent the entire time figuring out a way to unload you and make a run for the barn. His only redeeming quality was he'd run away with his head high and to the side so as not to break the dragging bridle reins.

While not our preference for a saddle horse, we did use him in other ways. He made a comical if not functional packhorse for our "kid" pack trips. We would cinch a packsaddle to him and tie to that all our treasures for the day that had been wrapped in an old green army blanket.

The mound on his back would be so huge it usually took a kid walking on each side holding the pack to keep it on the top side. Off we would go, lumbering up the road a mile or so to create our pretend world of cowboys and Indians and hunting camps.

In the height of his career, Mickey became the source of total indignation for my brother. Summer was irrigation time for all those hay meadows and Mickey was the assigned mode of transportation for the boy.

Shovel in hand, he would slip up on the pony bareback and head off for a day of directing water over hay fields. Dad told him he couldn't waste a good saddle horse on that job.

In the winter Mickey pulled a toboggan in the meadow for us – but only one direction. We'd jimmy rig some sort of harness for him and hook it to the sled.

Then we lead him to the top end of a long meadow, turned him towards home and let him go. Dependably, he would run as hard as he could back to the barn. It was always a wild ride.

None of us kids have ever forgotten Mickey Mouse. And none of us have owned a pony since then. Some lessons tend to take better than others.

A Room with a View Back Into Time

It could be the time of year. Christmas often brings a time of reflection for most people. Sometimes those reflections are not personal memories but a curious wondering of how it was for others "back then."

Yesterday I walked through a pasture tucked up against some steep rugged hills. Both the pasture and the hills lie in a pathway to a place and an era forgotten in time. The road that ribbons through both leads to what some would label a ghost town, White Oaks.

The pasture is on the 136-year-old Bar W Ranch that carries with it a history that includes ownership in 1869 by Lawrence G. Murphy of Lincoln County War fame and was later bought by its manager William C. McDonald, the man who became New Mexico's first governor.

It was in White Oaks in 1891, then a gold mining boomtown and one of the largest settled areas in the Territory of New Mexico, that McDonald came to work as a Deputy U.S. Metal Surveyor. He met and married Francis Tarbell McCourt and from that union came the linage of today's Bar W Ranch owners.

History rode through those hills because they are where they are. Standing in the late day sunshine, one has only to close their eyes to hear the hoof beats of riders and the rumbling of wagon wheels from more than a century ago.

My walk was a mission to photograph the last of the windmills standing on the Bar W before it makes way to more modern methods of pumping water. Near the well is a tiny adobe one-room house with crumbling walls and decaying wood floorboards, window and door frames. Next to it is a set falling-down-wood corrals that speaks of a time before welding machines and oil-field pipe built cattle working pens.

I could not stop myself from wanting the stories that humble home could tell me and to see the cattle that went through those gates with hardened Bar W cowboy crews at their heels.

Stepping gingerly inside the old building, I walked to the wall opposite the door and looked out the now glassless window frame. Still intact except for the panes, the window is hinged to open to the inside and offers a simple latch to keep it closed. Peering out, I saw only the hills beyond and the quiet still space that could yet be 1869.

The modest shack wore shelving on one wall that no doubt at one time held all the food stuffs, dishes and utensils. I thought of my pantry and wondered how it would be to have lived with such sparse offerings.

Looking back out the window, I tried to imagine Christmas in so basic surroundings. An image formed of a tiny tree standing in the corner and the light of a fire casting shadows on the walls. A cowboy was sitting in his one and only chair, leaning over a narrow table, reading his Bible by lantern light. He was reading the story of Christmas; the original story that began it all. That story told of a simple building in sparse times on a cold dark night.

As it was in the beginning, so shall we remember it today.

Away in the Manger
Out on the Range

As children, we learned about baby Jesus born in a manger and how the shepherds followed a star to where he lay.

Things haven't changed all that much out on the range. Christmas time, December in general, can be one of those times at the ranch when a fella wonders why he is having so much fun.

Like in the Bethlehem of baby Jesus' time, animals in winter need shelter in addition to food and water. Most ranches of any size can't put all the cows up in a barn when it's cold, so the momma cows will nest up next to a cedar tree in a canyon where the wind isn't as fierce.

Christmas morning will find the rancher, also referred to as the shepherd, tending his herd, otherwise known as the flock if he has sheep. No one mentioned to the cows it was a holiday and frankly, they aren't real concerned about it. They come running to the sound of the feed pickup like it was any other day.

So unlike the shepherds who followed the star, these shepherds are followed – calling to their flock with the sound of a honking horn. They will pour out the feed, get a good head count, break the ice on the water troughs and move on to the next pasture. If nothing has broke or escaped, he'll be home in time for Christmas dinner.

Ranching is a 24/7 kind of job. But no one keeps track of the hours, least of all the animals. Until this year, we were always calving out heifers (cows having their first babies) in December and not only were there the day time jobs, but always the every two hour labor ward checks during the night.

Crawling out of a warm bed in a warm house, to pull on warm clothes to wear into a cold night, walk around a pen full of cows with a flashlight to see if anyone is in need of assistance to birth their babies is not often the highlight of Western lore.

I might mention those cows would rather run over you than around you if you don't move fast enough. More than one discussion has been centered on the best kind of flashlight to use for a weapon.

Sometimes the laughs make it more worthwhile. Friend of mine made a 2 a.m. check on their heifers on 15-degree night. She found one in distress that would have to make a night trip to the vet for a cesarean.

My friend was in her usual night heifer check clothing of insulated coveralls pulled over her "slight" nightie. After about two hours in the very warm heated vet clinic it was obvious she was miserable. The vet never did understand why she wouldn't take her "coat" off. It's one of those moments not even Martha Stewart could doctor up.

So on Christmas morning when you are plodding to the coffee pot in your Poppa Moose slippers ready to lounge in leisure throughout the holiday morning, give a sacred thought of the cold night on which Jesus was born.

The reason for the season, as they say.

Then raise your coffee cup to the rancher who was out at daylight tending to your next serving of prime rib.

The Cowboy New Year

No, the calendar hasn't flipped over to the actual New Year as in "next year." For most people it is just the Fall season. But for the cowboy, fall is the beginning of a new year just like Monday is the beginning of a new week.

By the time November rolls around, most ranches in this part of the country have weaned their calves and/or shipped them off. They have had their one time payday for a year's hard work. One more year they have watched a cow buyer drive off down the dusty road in his big fancy car with his diamond pinky ring flashing in the sunlight. One more time they have let go a sigh of relief as the last cattle truck rolled over the cattle guard headed for feedlots and wheat pastures.

Yearling cattle operators have shipped the summer cattle and are looking to get the fall stockers received and tucked away in winter pastures.

Fall is when you get out all the jackets, down vests, wild rags and leggings. You make every effort to find the winter gloves, all of them, including the right and left one of each pair. It has been proven that while empty cardboard boxes multiply in captivity, winter gloves in matching pairs are an endangered species.

My first concession to the season is giving up my sandals for full-cover footwear. It usually doesn't happen before I've been seen in public several times wearing a turtleneck sweater and the aforementioned sandals.

The horses start getting long hair and spend more time at the feed bunk. They have little interest in working, socializing or doing anything but soaking up the afternoon sun.

Fall is when you start breaking the two year old colts and hope they turn out gentle enough so no one gets hurt. It is hard to maintain any cowboy athletic prowess with a bucking colt when you are dressed with enough layers to resemble the Michelin man.

Food changes from sandwiches and salads to pots of chili and a complete assortment of crock pot ready-to-eat cuisine options. Pumpkins are everywhere. Pumpkin cake and pumpkin bread are a favorite whether it is for the taste of cinnamon and clove or simply a good reason for the cream cheese frosting. While I happen to think pumpkin comes in a can, the real thing does look pretty sitting around next to Indian corn or bundled corn stocks.

It is not yet calving season and there is not yet any ice to break on the water tanks. The feed pickups stand by ready for work. But the season is too short to start any major fencing, pipelining or corral building. It is not that winter will be idle but winter has a specific set of jobs which mostly leave no time for special projects.

Fall is the time to review what has been accomplished during the year – you can't get it back but you can always hope to improve on it. Ranching is like that. You always look forward to getting this year over with so you can start on the next one. Ranchers just get to start a little earlier with their New Year resolutions.

Happy New Year

Valentine's Day...for Romance or Economics

Valentine's Day is looming on the near horizon and men of the west are restless with worry. The pressure is on.

Getting to the day and through the day in the good graces of the little purty is high priority for a few but not all.

One veteran cowboy when asked what he thought about Valentine's Day replied, "Not much. I don't think about it at all. You don't want to get that started – birthdays, Valentine's and all those holidays. If you never start paying attention to them, then the Little Woman never expects it."

When discussing Valentine's Day, you will often hear "Where did that get started anyway?"

Some will attest to the theory that the day was invented as a clever ploy to stimulate the economy in an otherwise sluggish time of year. Greeting card companies, florists, jewelers and chocolate manufacturers who flourish because of it would have to agree.

Although shrouded in mystery, history records a couple of theories about this annual homage paid to the patron saint of the day, St. Valentine.

One legend contends that Valentine was a priest who served during the third century in Rome. When Emperor Claudius II decided that single men made better soldiers than those with wives and families, he outlawed marriage for young men that were his crop of potential soldiers.

Valentine, realizing the injustice of the decree, defied Claudius and continued to perform marriages in secret for young lovers. When

Valentine's actions were discovered, Claudius ordered that he be put to death.

According to another legend, Valentine actually sent the first "Valentine" greeting himself. While in prison awaiting his execution, it is believed that Valentine fell in love with a young girl who may have been his jailor's daughter who visited him during his confinement. Before his death, it is alleged that he wrote her a letter, which he signed "From your Valentine," an expression that is still in use today.

Although the truth behind the Valentine legends is murky, the stories certainly emphasize his appeal as a sympathetic, heroic, and, most importantly, romantic figure. It's no surprise that by the Middle Ages, Valentine was one of the most popular saints in England and France.

Valentine's Day advertisements, even locally in a rural part of the world, promise evenings of lasting romance and adoration if you will just come dine with them for only $175 a couple. I don't foresee too many pickup trucks leaving the ranch for that one.

There will be some "romantic" gestures made at those dirt road residences. It may not be wine and roses but a cowboy on a Valentine's Day date will offer a romantic late night walk through the frosty pastures for a "just once more" check of the cows. After all, it is calving season.

My dad, while waiting on a hay buyer, took advantage of the available material and created a baling wire heart complete with the arrow through it for the one time ever Valentine's gift for my mother.

I got a Valentine once. The card was in Spanish because that is all that was left at the store and I'm not sure that it wasn't a birthday card. With it came a box of chocolates, his favorite kind, since he knew I was holding fast to my diet.

Valentine's Day may not be just a scheduled boost to the February economy. It can also be a right-on-target guaranteed successful plot to end any New Year's diet resolutions. It is one more reason not to make that Jan. 1 promise. Wait until Feb. 15 and don't share your chocolate.

Photos by Julie Carter

Page 13 – Julie has been photo shopped onto a Julie Carter photo of a set of picket corrals and windmill west of Carrizozo, NM on the Grant Kinzer Ranch.

Page 22 – Spring brings the blooms of the yucca to Lincoln County, NM.

Page 34 – Cow skull on a stucco wall. One of Julie's favorite things to collect is the skulls of horned cattle.

Page 48 – This room of a barn, now holding saddles, once served as a homestead home to ranchers near Ancho, NM.

Page 51 – Twisted wire on an old cedar fence post. Bar W Ranch, Carrizozo, NM

Page 54 – Bar W Ranch windmill, Carrizozo, NM.

Page 56 – Local rancher's flatbed pickup parked at the post office at Carrizozo, NM.

Page 65 – Old corrals on Nogal Mesa near Nogal, NM.

Page 72 – Washed with golden light of a sunset, this old loading chute is a reminder of days gone by. Bar W Ranch, Carrizozo, NM.

Page 84 – Windmill at sunset at the Bar W Ranch, Carrizozo, NM.

Page 97 – An old outhouse at the Bar W Ranch, Carrizozo, NM.

Page 104 – Old things always mark old places and sometimes whisper of the old days. Bar W Ranch, Carrizozo, NM.

Page 107 – Joe Smith, a local New Mexico saddle maker, has a wall full of completed headstalls and breast collars.

Page 121 – Old wooden squeeze chutes stand as relics next to the mercantile in Mountainair, NM.

Page 124 – Skulls of horned cattle stand as symbols of the plains, a favorite of Western artists and a favorite for Julie's camera.

Page 127 – Brightly colored spur straps and a bell on a bull rope with a cross on it called to Julie's camera at the Lincoln County Rodeo Club event in Capitan, NM.

Page 130 – A sight that could be almost anywhere in any cowboy's corral. This one happened to be at the Wilson Ranch, Ancho, NM.

Page 144 – Weathered logs of an old barn still in use. Wilson Ranch, Ancho, NM.

Page 149 – This old truck sits "parked" in a field near Claunch, NM.

Page 158 – Sweat-soaked Stetson.

Page 167 – Some stack them long, some stack them high. Lincoln County, NM ranch tack room.

Page 181 – Carrizo Springs, Bar W Ranch, Carrizozo, NM.

Page 186 – Atkinson-Lindsey Ranch windmill, Claunch, NM.

Page 196 – Lincoln County, NM is cattle country.

Page 206 – And the buzzards return to Carrizozo like the swallows to Capistrano.

ABOUT THE AUTHOR

Julie Carter paints cowboy pictures with wit and words in her attempt to bring a unique point of view to cowboy life - be it cowgirl woes, cowboy behavior, rodeo adventures or simple essays about the nature of the enigma called "cowboy."

From her ranch raising in the mountains of southern Colorado to her ranch living in New Mexico and hundreds of rodeo arenas in between, she writes with the intent to make you laugh, make you think or a little of both.

The stories are those she has experienced herself or were told to her by others. The lifestyle and experiences are not unique to her, but Julie writes as the voice of those that came before her and those that still live down dirt roads far behind the cattle guard.

Julie is a fulltime staff writer for the *Ruidoso News* in Lincoln County, New Mexico, and writes a weekly syndicated column, 'Cowgirl Sass and Savvy.'

Visit Julie at her website: julie-carter.com

or contact her at: julie@julie-carter.com

www.ingramcontent.com/pod-product-compliance
Lightning Source LLC
Chambersburg PA
CBHW051825090426
42736CB00011B/1651